Machine Learning Infrastructure and Best Practices for Software Engineers

Take your machine learning software from a prototype to a fully fledged software system

Miroslaw Staron

Machine Learning Infrastructure and Best Practices for Software Engineers

Group Product Manager: Niranjan Naikwadi

Publishing Product Manager: Yasir Ali Khan

Book Project Manager: Hemangi Lotlikar

Senior Editor: Sushma Reddy

Technical Editor: Kavyashree K S

Copy Editor: Safis Editing

Proofreader: Safis Editing

Indexer: Hemangini Bari

Production Designer: Gokul Raj S.T

DevRel Marketing Coordinator: Vinishka Kalra

First published: January 2024

Production reference: 1170124

Published by
Packt Publishing Ltd.
Grosvenor House
11 St Paul's Square
Birmingham
B3 1RB, UK

ISBN 978-1-83763-406-4

www.packtpub.com

Writing a book with a lot of practical examples requires a lot of extra time, which is often taken from family and friends. I dedicate this book to my family – Alexander, Cornelia, Viktoria, and Sylwia – who always supported and encouraged me, and to my parents and parents-in-law, who shaped me to be who I am.

– Miroslaw Staron

Contributors

About the author

Miroslaw Staron is a professor of Applied IT at the University of Gothenburg in Sweden with a focus on empirical software engineering, measurement, and machine learning. He is currently editor-in-chief of *Information and Software Technology* and co-editor of the regular *Practitioner's Digest* column of IEEE Software. He has authored books on automotive software architectures, software measurement, and action research. He also leads several projects in AI for software engineering and leads an AI and digitalization theme at Software Center. He has written over 200 journal and conference articles.

I would like to thank my family for their support in writing this book. I would also like to thank my colleagues from the Software Center program who provided me with the ability to develop my ideas and knowledge in this area – in particular, Wilhelm Meding, Jan Bosch, Ola Söder, Gert Frost, Martin Kitchen, Niels Jørgen Strøm, and several other colleagues. One person who really ignited my interest in this area is of course Mirosław "Mirek" Ochodek, to whom I am extremely grateful. I would also like to thank the funders of my research, who supported my studies throughout the years. I would like to thank my Ph.D. students, who challenged me and encouraged me to always dig deeper into the topics. I'm also very grateful to the reviewers of this book – Hongyi Zhang and Sushant K. Pandey, who provided invaluable comments and feedback for the book. Finally, I would like to extend my gratitude to my publishing team – Hemangi Lotlikar, Sushma Reddy, and Anant Jaint – this book would not have materialized without you!

About the reviewers

Hongyi Zhang is a researcher at Chalmers University of Technology with over five years of experience in the fields of machine learning and software engineering. Specializing in machine learning, edge/cloud computing, and software engineering, his research merges machine learning theory and software applications, driving tangible improvements in industrial machine learning ecosystems.

Sushant Kumar Pandey is a dedicated post-doctoral researcher at the Department of CSE, Chalmers at the University of Gothenburg, Sweden, who seamlessly integrates academia with industry, collaborating with Volvo Cars in Gothenburg. Armed with a Ph.D. in CSE from the esteemed Indian Institute of Technology (BHU), India, Sushant specializes in the application of AI in software engineering. His research advances technology's transformative potential. As a respected reviewer for prestigious venues such as IST, KBS, EASE, and ESWA, Sushant actively contributes to shaping the discourse in his field. Beyond research, he leverages his expertise to mentor students, fostering innovation and excellence in the next generation of professionals.

Table of Contents

Part 1: Machine Learning Landscape in Software Engineering

1

2

Part 2: Data Acquisition and Management

6

7

8

Part 3: Design and Development of ML Systems

9

10

11

12

Designing Machine Learning Pipelines (MLOps) and Their Testing   221

13

Designing and Implementing Large-Scale, Robust ML Software   241

Part 4: Ethical Aspects of Data Management and ML System Development

14

Ethics in Data Acquisition and Management   265

15

Ethics in Machine Learning Systems   279

16

Integrating ML Systems in Ecosystems   289

17

Summary and Where to Go Next   305

Index   315

Other Books You May Enjoy   322

Preface

Machine learning has gained a lot of popularity in recent years. The introduction of large language models such as GPT-3 and 4 only increased the speed of the development of this field. These large language models have become so powerful that it is almost impossible to train them on a local computer. However, this is not necessary at all. These language models provide the ability to create new tools without the need to train them because they can be steered by the context window and the prompt.

In this book, my goal is to show how machine learning models can be trained, evaluated, and tested – both in the context of a small prototype and in the context of a fully-fledged software product. The primary objective of this book is to bridge the gap between theoretical knowledge and practical implementation of machine learning in software engineering. It aims to equip you with the skills necessary to not only understand but also effectively implement and innovate with AI and machine learning technologies in your professional pursuits.

The journey of integrating machine learning into software engineering is as thrilling as it is challenging. As we delve into the intricacies of machine learning infrastructure, this book serves as a comprehensive guide, navigating through the complexities and best practices that are pivotal for software engineers. It is designed to bridge the gap between the theoretical aspects of machine learning and the practical challenges faced during implementation in real-world scenarios.

We begin by exploring the fundamental concepts of machine learning, providing a solid foundation for those new to the field. As we progress, the focus shifts to the infrastructure – the backbone of any successful machine learning project. From data collection and processing to model training and deployment, each step is crucial and requires careful consideration and planning.

A significant portion of the book is dedicated to best practices. These practices are not just theoretical guidelines but are derived from real-life experiences and case studies that my research team discovered during our work in this field. These best practices offer invaluable insights into handling common pitfalls and ensuring the scalability, reliability, and efficiency of machine learning systems.

Furthermore, we delve into the ethics of data and machine learning algorithms. We explore the theories behind ethics in machine learning, look closer into the licensing of data and models, and finally, explore the practical frameworks that can quantify bias in data and models in machine learning.

This book is not just a technical guide; it is a journey through the evolving landscape of machine learning in software engineering. Whether you are a novice eager to learn, or a seasoned professional seeking to enhance your skills, this book aims to be a valuable resource, providing clarity and direction in the exciting and ever-changing world of machine learning.

Who this book is for

This book is meticulously crafted for software engineers, computer scientists, and programmers who seek practical applications of artificial intelligence and machine learning in their field. The content is tailored to impart foundational knowledge on working with machine learning models, viewed through the lens of a programmer and system architect.

The book presupposes familiarity with programming principles, but it does not demand expertise in mathematics or statistics. This approach ensures accessibility to a broader range of professionals and enthusiasts in the software development domain. For those of you without prior experience in Python, this book necessitates acquiring a basic understanding of the language. However, the material is structured to facilitate a rapid and comprehensive grasp of Python essentials. Conversely, for those proficient in Python but not yet seasoned in professional programming, this book serves as a valuable resource for transitioning into the realm of software engineering with a focus on AI and ML applications.

What this book covers

Chapter 1, *Machine Learning Compared to Traditional Software*, explores where these two types of software systems are most appropriate. We learn about the software development processes that programmers use to create both types of software and we also learn about the classical four types of machine learning software – rule-based, supervised, unsupervised, and reinforcement learning. Finally, we also learn about the different roles of data in traditional and machine learning software.

Chapter 2, *Elements of a Machine Learning System*, reviews each element of a professional machine learning system. We start by understanding which elements are important and why. Then, we explore how to create such elements and how to work by putting them together into a single machine learning system – the so-called machine learning pipeline.

Chapter 3, *Data in Software Systems – Text, Images, Code, and Features*, introduces three data types – images, texts, and formatted text (program source code). We explore how each of these types of data can be used in machine learning, how they should be annotated, and for what purpose. Introducing these three types of data provides us with the possibility to explore different ways of annotating these sources of data.

Chapter 4, *Data Acquisition, Data Quality, and Noise*, dives deeper into topics related to data quality. We go through a theoretical model for assessing data quality and we provide methods and tools to operationalize it. We also look into the concept of noise in machine learning and how to reduce it by using different tokenization methods.

Chapter 5, *Quantifying and Improving Data Properties*, dives deeper into the properties of data and how to improve them. In contrast to the previous chapter, we work on feature vectors rather than raw data. The feature vectors are already a transformation of the data; therefore, we can change such properties as noise or even change how the data is perceived. We focus on the processing of text, which is an important part of many machine learning algorithms nowadays. We start by understanding how to transform data into feature vectors using simple algorithms, such as bag of words, so that we can work on feature vectors.

Chapter 6, Processing Data in Machine Learning Systems, dives deeper into the ways in which data and algorithms are entangled. We talk a lot about data in generic terms, but in this chapter, we explain what kind of data is needed in machine learning systems. We explain the fact that all kinds of data are used in numerical form – either as a feature vector or as more complex feature matrices. Then, we will explain the need to transform unstructured data (e.g., text) into structured data. This chapter will lay the foundations for going deeper into each type of data, which is the content of the next few chapters.

Chapter 7, Feature Engineering for Numerical and Image Data, focuses on the feature engineering process for numerical and image data. We start by going through the typical methods such as **Principal Component Analysis** (**PCA**), which we used previously for visualization. We then move on to more advanced methods such as the **t-Student Distribution Stochastic Network Embeddings** (**t-SNE**) and **Independent Component Analysis** (**ICA**). What we end up with is the use of autoencoders as a dimensionality reduction technique for both numerical and image data.

Chapter 8, Feature Engineering for Natural Language Data, explores the first steps that made the transformer (GPT) technologies so powerful – feature extraction from natural language data. Natural language is a special kind of data source in software engineering. With the introduction of GitHub Copilot and ChatGPT, it became evident that machine learning and artificial intelligence tools for software engineering tasks are no longer science fiction.

Chapter 9, Types of Machine Learning Systems – Feature-Based and Raw Data-Based (Deep Learning), explores different types of machine learning systems. We start from classical machine learning models such as random forest and we move on to convolutional and GPT models, which are called deep learning models. Their name comes from the fact that they use raw data as input and the first layers of the models include feature extraction layers. They are also designed to progressively learn more abstract features as the input data moves through these models. This chapter demonstrates each of these types of models and progresses from classical machine learning to the generative AI models.

Chapter 10, Training and Evaluation of Classical ML Systems and Neural Networks, goes a bit deeper into the process of training and evaluation. We start with the basic theory behind different algorithms and then we show how they are trained. We start with the classical machine learning models, exemplified by the decision trees. Then, we gradually move toward deep learning where we explore both the dense neural networks and some more advanced types of networks.

Chapter 11, Training and Evaluation of Advanced ML Algorithms – GPT and Autoencoders, explores how generative AI models work based on GPT and Bidirectional Encoder Representation Transformers (BERT). These models are designed to generate new data based on the patterns that they were trained on. We also look at the concept of autoencoders, where we train an autoencoder to generate new images based on the previously trained data.

Chapter 12, Designing Machine Learning Pipelines and their Testing, describes how the main goal of MLOps is to bridge the gap between data science and operations teams, fostering collaboration and ensuring that machine learning projects can be effectively and reliably deployed at scale. MLOps helps to automate and optimize the entire machine learning life cycle, from model development to deployment and maintenance, thus improving the efficiency and effectiveness of ML systems in production. In this chapter, we learn how machine learning systems are designed and operated in practice. The chapter shows how pipelines are turned into a software system, with a focus on testing ML pipelines and their deployment at Hugging Face.

Chapter 13, Designing and Implementation of Large-Scale, Robust ML Software, explains how to integrate the machine learning model with a graphical user interface programmed in Gradio and storage in a database. We use two examples of machine learning pipelines – an example of the model for predicting defects from our previous chapters and a generative AI model to create pictures from a natural language prompt.

Chapter 14, Ethics in Data Acquisition and Management, starts by exploring a few examples of unethical systems that show bias, such as credit ranking systems that penalize certain minorities. We also explain the problems with using open source data and revealing the identities of subjects. The core of the chapter, however, is the explanation and discussion on ethical frameworks for data management and software systems, including the IEEE and ACM codes of conduct.

Chapter 15, Ethics in Machine Learning Systems, focuses on the bias in machine learning systems. We start by exploring sources of bias and briefly discussing these sources. We then explore ways to spot biases, how to minimize them, and finally, how to communicate potential biases to the users of our system.

Chapter 16, Integration of ML Systems in Ecosystems, explains how packaging the ML systems into web services allows us to integrate them into workflows in a very flexible way. Instead of compiling or using dynamically linked libraries, we can deploy machine learning components that communicate over HTTP protocols using JSON protocols. In fact, we have already seen how to use that protocol by using the GPT-3 model that is hosted by OpenAI. In this chapter, we explore the possibility of creating our own Docker container with a pre-trained machine learning model, deploying it, and integrating it with other components.

Chapter 17, Summary and Where to Go Next, revisits all the best practices and summarizes them per chapter. In addition, we also look into what the future of machine learning and AI may bring to software engineering.

To get the most out of this book

In this book, we use Python and PyTorch, so you need to have these two installed on your system. I used them on Windows and Linux, but they can also be used in cloud environments such as Google Colab or GitHub Codespaces (both were tested).

Software/hardware covered in the book	Operating system requirements
Python 3.11	Windows, Ubuntu, Debian Linux, or Windows Subsystem for Linux (WSL)
PyTorch 2.1	Windows, Ubuntu, or Debian Linux

If you are using the digital version of this book, we advise you to type the code yourself or access the code from the book's GitHub repository (a link is available in the next section). Doing so will help you avoid any potential errors related to the copying and pasting of code.

Download the example code files

You can download the example code files for this book from GitHub at https://github.com/ PacktPublishing/Machine-Learning-Infrastructure-and-Best-Practices-for-Software-Engineers. If there's an update to the code, it will be updated in the GitHub repository.

We also have other code bundles from our rich catalog of books and videos available at https:// github.com/PacktPublishing/. Check them out!

Conventions used

There are a number of text conventions used throughout this book.

`Code in text`: Indicates code words in text, database table names, folder names, filenames, file extensions, pathnames, dummy URLs, user input, and Twitter handles. Here is an example: "The model itself is created one line above, in the model = LinearRegression() line."

A block of code is set as follows:

```
def fibRec(n):

    if n < 2:

        return n

    else:

        return fibRec(n-1) + fibRec(n-2)
```

Any command-line input or output is written as follows:

```
>python app.py
```

> **Best practices**
> Appear like this.

Get in touch

Feedback from our readers is always welcome.

General feedback: If you have questions about any aspect of this book, email us at customercare@ packtpub.com and mention the book title in the subject of your message.

Errata: Although we have taken every care to ensure the accuracy of our content, mistakes do happen. If you have found a mistake in this book, we would be grateful if you would report this to us. Please visit www.packtpub.com/support/errata and fill in the form.

Piracy: If you come across any illegal copies of our works in any form on the internet, we would be grateful if you would provide us with the location address or website name. Please contact us at copyright@packt.com with a link to the material.

If you are interested in becoming an author: If there is a topic that you have expertise in and you are interested in either writing or contributing to a book, please visit authors.packtpub.com.

Share Your Thoughts

Once you've read *Machine Learning Infrastructure and Best Practices for Software Engineers*, we'd love to hear your thoughts! Scan the QR code below to go straight to the Amazon review page for this book and share your feedback.

https://packt.link/r/1-837-63406-8

Your review is important to us and the tech community and will help us make sure we're delivering excellent quality content.

Download a free PDF copy of this book

Thanks for purchasing this book!

Do you like to read on the go but are unable to carry your print books everywhere?

Is your eBook purchase not compatible with the device of your choice?

Don't worry, now with every Packt book you get a DRM-free PDF version of that book at no cost.

Read anywhere, any place, on any device. Search, copy, and paste code from your favorite technical books directly into your application.

The perks don't stop there, you can get exclusive access to discounts, newsletters, and great free content in your inbox daily

Follow these simple steps to get the benefits:

1. Scan the QR code or visit the link below

https://packt.link/free-ebook/978-1-83763-406-4

2. Submit your proof of purchase
3. That's it! We'll send your free PDF and other benefits to your email directly

Part 1: Machine Learning Landscape in Software Engineering

Traditionally, Machine Learning (ML) was considered to be a niche domain in software engineering. No large software systems used statistical learning in production. This changed in the 2010s when recommendation systems started to utilize large quantities of data – for example, to recommend movies, books, or music. With the rise of transformer technologies, this has changed. Commonly known products such as ChatGPT popularized these techniques and showed that they are no longer niche products, but have entered the mainstream software products and services. Software engineering needs to keep up and we need to know how to create the software based on these modern machine learning models. In this first part of the book, we look at how machine learning changes software development and how we need to adapt to these changes.

This part has the following chapters:

- *Chapter 1, Machine Learning Compared to Traditional Software*
- *Chapter 2, Elements of a Machine Learning System*
- *Chapter 3, Data in Software Systems – Text, Images, Code, and Features*
- *Chapter 4, Data Acquisition, Data Quality, and Noise*
- *Chapter 5, Quantifying and Improving Data Properties*

1

Machine Learning Compared to Traditional Software

Machine learning software is a special kind of software that finds patterns in data, learns from them, and even recreates these patterns on new data. Developing the machine learning software is, therefore, focused on finding the right data, matching it with the appropriate algorithm, and evaluating its performance. Traditional software, on the contrary, is developed with the algorithm in mind. Based on software requirements, programmers develop algorithms that solve specific tasks and then test them. Data is secondary, although not completely unimportant. Both types of software can co-exist in the same software system, but the programmer must ensure compatibility between them.

In this chapter, we'll explore where these two types of software systems are most appropriate. We'll learn about the software development processes that programmers use to create both types of software. We'll also learn about the four classical types of machine learning software – rule-based learning, supervised learning, unsupervised learning, and reinforcement learning. Finally, we'll learn about the different roles of data in traditional and machine learning software – as input to pre-programmed algorithms in traditional software and input to training models in machine learning software.

The best practices introduced in this chapter provide practical guidance on when to choose each type of software and how to assess the advantages and disadvantages of these types. By exploring a few modern examples, we'll understand how to create an entire software system with machine learning algorithms at the center.

In this chapter, we're going to cover the following main topics:

- Machine learning is not a traditional software

- Probability and software – how well do they go together?

- Testing and validation – the same but different

Machine learning is not traditional software

Although machine learning and artificial intelligence have been around since the 1950s, introduced by Alan Turing, they only became popular with the first MYCIN system and our understanding of machine learning systems changed over time. It was not until the 2010s that we started to perceive, design, and develop machine learning in the same way as we do today (in 2023). In my view, two pivotal moments shaped the landscape of machine learning as we see it today.

The first pivotal moment was the focus on big data in the late 2000s and early 2010s. With the introduction of smartphones, companies started to collect and process increasingly large quantities of data, mostly about our behavior online. One of the companies that perfected this was Google, which collected data about our searches, online behavior, and usage of Google's operating system, Android. As the volume of the collected data increased (and its speed/velocity), so did its value and the need for its veracity – the five Vs. These five Vs – volume, velocity, value, veracity, and variety – required a new approach to working with data. The classical approach of relational databases (SQL) was no longer sufficient. Relational databases became too slow in handling high-velocity data streams, which gave way to map-reduce algorithms, distributed databases, and in-memory databases. The classical approach of relational schemas became too constraining for the variety of data, which gave way for non-SQL databases, which stored documents.

The second pivotal moment was the rise of modern machine learning algorithms – deep learning. Deep learning algorithms are designed to handle unstructured data such as text, images, or music (compared to structured data in the form of tables and matrices). Classical machine learning algorithms, such as regression, decision trees, or random forest, require data in a tabular form. Each row is a data point, and each column is one characteristic of it – a feature. The classical models are designed to handle relatively small datasets. Deep learning algorithms, on the other hand, can handle large datasets and find more complex patterns in the data because of the power of large neural networks and their complex architectures.

Machine learning is sometimes called *statistical learning* as it is based on statistical methods. The statistical methods calculate properties of data (such as mean values, standard deviations, and coefficients) and thus find patterns in the data. The core characteristic of machine learning is that it uses data to find patterns, learn from them, and then repeat these patterns on new data. We call this way of learning patterns training, and repeating these patterns as reasoning, or in machine learning language, *predicting*. The main benefits of using machine learning software come from the fact that we do not need to design the algorithms – we focus on the problem to be solved and the data that we use to solve the problem. *Figure 1.1* shows an example of how such a flowchart of machine learning software can be realized.

First, we import a generic machine learning model from a library. This generic model has all elements that are specific to it, but it is not trained to solve any tasks. An example of such a model is a decision tree model, which is designed to learn dependencies in data in the form of decisions (or data splits), which it uses later for new data. To make this model somewhat useful, we need to train it. For that, we need data, which we call the training data.

Second, we evaluate the trained model on new data, which we call the test data. The evaluation process uses the trained model and applies it to check whether its inferences are correct. To be precise, it checks to which degree the inferences are correct. The training data is in the same format as the test data, but the content of these datasets is different. No data point should be present in both.

In the third step, we use the model as part of a software system. We develop other non-machine learning components, and we connect them to the trained model. The entire software system usually consists of data procurement components, real-time validation components, data cleaning components, user interfaces, and business logic components. All these components, including the machine learning model, provide a specific functionality for the end user. Once the software system has been developed, it needs to be tested, which is where the input data comes into play. The input data is something that the end user inputs to the system, such as by filling in a form. The input data is designed in such a way that has both the input and expected output – to test whether the software system works correctly.

Finally, the last step is to deploy the entire system. The deployment can be very different, but most modern machine learning systems are organized into two parts – the onboard/edge algorithms for non-machine learning components and the user interface, and the offboard/cloud algorithms for machine learning inferences. Although it is possible to deploy all parts of the system on the target device (both machine learning and non-machine learning components), complex machine learning models require significant computational power for good performance and seamless user experience. The principle is simple – more data/complex data means more complex models, which means that more computational power is needed:

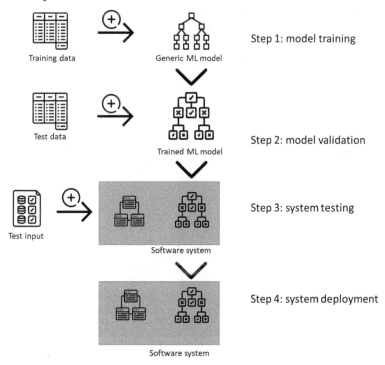

Figure 1.1 – Typical flow of machine learning software development

As shown in *Figure 1.1*, one of the crucial elements of the machine learning software is the *model*, which is one of the generic machine learning models, such as a neural network, that's been trained on specific data. Such a model is used to make predictions and inferences. In most systems, this kind of component – the model – is often prototyped and developed in Python.

Models are trained for different datasets and, therefore, the core characteristic of machine learning software is its dependence on that dataset. An example of such a model is a vision system, where we train a machine learning algorithm such as a **convolutional neural network** (**CNN**) to classify images of cats and dogs.

Since the models are trained on specific datasets, they perform best on similar datasets when making inferences. For example, if we train a model to recognize cats and dogs in 160 x 160-pixel grayscale images, the model can recognize cats and dogs in such images. However, the same model will perform very poorly (if at all!) if it needs to recognize cats and dogs in colorful images instead of grayscale images – the accuracy of the classification will be low (close to 0).

On the other hand, when we develop and design traditional software systems, we do not rely on data that much, as shown in *Figure 1.2*. This figure provides an overview of a software development process for traditional, non-machine learning software. Although it is depicted as a flow, it is usually an iterative process where *Steps 1* to *3* are done in cycles, each one ending with new functionality added to the product.

The first step is developing the software system. This includes the development of all its components – user interface, business logic (processing), handling of data, and communication. The step does not involve much data unless the software engineer creates data for testing purposes.

The second step is system testing, where we use input data to validate the software system. In essence, this step is almost identical to testing machine learning software. The input data is complemented with the expected outcome data, which allows software testers to assess whether the software works correctly.

The third step is to deploy the software. The deployment can be done in many ways. However, if we consider traditional software that is similar in function to machine learning software, it is usually simpler. It usually does not require deployment on the cloud, just like machine learning models:

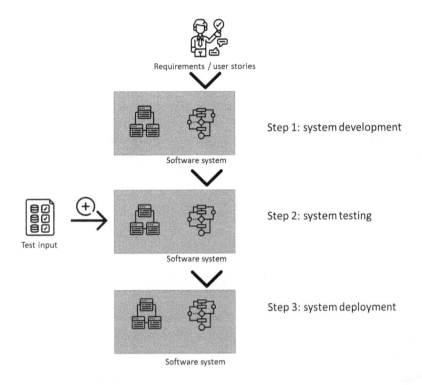

Figure 1.2 - Typical flow of traditional software development

The main difference between traditional software and machine learning-based software is that we need to design, develop, and test all the elements of the traditional software. In machine learning-based software, we take an empty model, which contains all the necessary elements, and we use the data to train it. We do not need to develop the individual components of the machine learning model from scratch.

One of the main parts of traditional software is the *algorithm*, which is developed by software engineers from scratch, based on the requirements or user stories. The algorithm is usually written as a sequential set of steps that are implemented in a programming language. Naturally, all algorithms use data to operate on it, but they do it differently than machine learning systems. They do it based on the software engineer's design – *if x, then y* or something similar.

We usually consider these traditional algorithms as deterministic, explainable, and traceable. This means that the software engineer's design decisions are documented in the algorithm and the algorithm can be analyzed afterward. They are deterministic because they are programmed based on rules; there is no training from data or identifying patterns from data. They are explainable because they are designed by programmers and each line of the program has a predefined meaning. Finally, they are traceable as we can debug every step of these programs.

However, there is a drawback – the software engineer needs to thoroughly consider all corner cases and understand the problem very well. The data that the software engineer uses is only to support them in analyzing the algorithm, not training it.

An example of a system that can be implemented using both machine learning algorithms and traditional ones is one for reading passport information. Instead of using machine learning for image recognition, the software uses specific marks in the passport (usually the <<< sequence of characters) to mark the beginning of the line or the beginning of the sequence of characters denoting a surname. These marks can be recognized quite quickly using rule-based **optical character recognition (OCR)** algorithms without the need for deep learning or CNNs.

Therefore, I would like to introduce the first best practice.

Best practice #1

Use machine learning algorithms when your problem is focused on data, not on the algorithm.

When selecting the right technology, we need to understand whether it is based on the classical approach, where the design of the algorithm is in focus, or whether we need to focus on handling data and finding patterns in it. It is usually beneficial to start with the following guidelines.

If the problem requires processing large quantities of data in raw format, use the machine learning approach. Examples of such systems are conversational bots, image recognition tools, text processing tools, or even prediction systems.

However, if the problem requires traceability and control, use the traditional approach. Examples of such systems are control software in cars (anti-lock braking, engine control, and so on) and embedded systems.

If the problem requires new data to be generated based on the existing data, a process known as *data manipulation*, use the machine learning approach. Examples of such systems are image manipulation programs (DALL-E), text generation programs, deep fake programs, and source code generation programs (GitHub Copilot).

If the problem requires adaptation over time and optimization, use machine learning software. Examples of such systems are power grid optimization software, non-playable character behavior components in computer games, playlist recommendation systems, and even GPS navigation systems in modern cars.

However, if the problem requires stability and traceability, use the traditional approach. Examples of such systems are systems to make diagnoses and recommendation systems in medicine, safety-critical systems in cars, planes, and trains, and infrastructure controlling and monitoring systems.

Supervised, unsupervised, and reinforcement learning – it is just the beginning

Now is a good time to mention that the field of machine learning is huge, and it is organized into three main areas – supervised learning, unsupervised learning, and reinforcement learning. Each of these areas has hundreds of different algorithms. For example, the area of supervised learning has over 1,000 algorithms, all of which can be automatically selected by meta-heuristic algorithms such as AutoML:

- **Supervised learning**: This is a group of algorithms that are trained based on annotated data. The data that's used in these algorithms needs to have a *target* or a *label*. The label is used to tell the algorithm which pattern to look for. For example, such a label can be *cat* or *dog* for each image that the supervised learning model needs to recognize. Historically, supervised learning algorithms are the oldest ones as they come directly from statistical methods such as linear regression and multinomial regression. Modern algorithms are advanced and include methods such as deep learning neural networks, which can recognize objects in 3D images and segment them accordingly. The most advanced algorithms in this area are deep learning and multimodal models, which can process text and images at the same time.

  A sub-group of supervised learning algorithms is **self-supervised models**, which are often based on transformer architectures. These models do not require labels in the data, but they use the data itself as labels. The most prominent examples of these algorithms are translation models for natural languages and generative models for images or texts. Such algorithms are trained by masking words in the original texts and predicting them. For the generative models, these algorithms are trained by masking parts of their output to predict it.

- **Unsupervised learning**: This is a group of models that are applied to find patterns in data without any labels. These models are not trained, but they use statistical properties of the input data to find patterns. Examples of such algorithms are clustering algorithms and semantic map algorithms. The input data for these algorithms is not labeled and the goal of applying these algorithms is to find structure in the dataset according to similarities; these structures can then be used to add labels to this data. We encounter these algorithms daily when we get recommendations for products to buy, books to read, music to listen to, or films to watch.

- **Reinforcement learning**: This is a group of models that are applied to data to solve a particular task given a goal. For these models, we need to provide this goal in addition to the data. It is called the *reward function,* and it is an expression that defines when we achieve the goal. The model is trained based on this fitness function. Examples of such models are algorithms that play Go, Chess, or StarCraft. These algorithms are also used to solve hard programming problems (AlphaCode) or optimize energy consumption.

So, let me introduce the second best practice.

> **Best practice #2**
>
> Before you start developing a machine learning system, do due diligence and identify the right group of algorithms to use.

As each of these groups of models has different characteristics, solves different problems, and requires different data, a mistake in selecting the right algorithm can be costly. Supervised models are very good at solving problems related to predictions and classifications. The most powerful models in this area can compete with humans in selected areas – for example, GitHub Copilot can create programs that can pass as human-written. Unsupervised models are very powerful if we want to group entities and make recommendations. Finally, reinforcement learning models are the best when we want to have continuous optimization with the need to retrain models every time the data or the environment changes.

Although all these models are based on statistical learning, they are all components of larger systems to make them useful. Therefore, we need to understand how this probabilistic and statistical nature of machine learning goes with traditional, digital software products.

An example of traditional and machine learning software

To illustrate the difference between traditional software and machine learning software, let's implement the same program using these two paradigms. We'll implement a program that calculates a Fibonacci sequence using the traditional approach, which we have seen a million times in computer science courses. Then, we'll implement the same program using machine learning models – or one model to be exact – that is, logistic regression.

The traditional implementation is presented here. It is based on one recursive function and a loop that tests it:

```
# a recursive function to calculate the fibonacci number
# this is a standard solution that is used in almost all
# of computer science examples
def fibRec(n):
  if n < 2:
      return n
  else:
      return fibRec(n-1) + fibRec(n-2)

# a short loop that uses the above function
for i in range(23):
  print(fibRec(i))
```

The implementation is very simple and is based on the algorithm – in our case, the fibRec function. It is simplistic, but it has its limitations. The first one is its recursive implementation, which costs resources. Although it can be written as an iterative one, it still suffers from the second problem – it is focused on the calculations and not on the data.

Now, let's see how the machine learning implementation is done. I'll explain this by dividing it into two parts – data preparation and model training/inference:

```
#predicting fibonacci with linear regression
import pandas as pd
import numpy as np
from sklearn.linear_model import LinearRegression


# training data for the algorithm
# the first two columns are the numbers and the third column is the
result
dfTrain = pd.DataFrame([[1, 1, 2],
                        [2, 1, 3],
                        [3, 2, 5],
                        [5, 3, 8],
                        [8, 5, 13]
])


# now, let's make some predictions
# we start the sequence as a list with the first two numbers
lstSequence = [0,1]

# we add the names of the columns to make it look better
dfTrain.columns = ['first number','second number','result']
```

In the case of machine learning software, we prepare data to train the algorithm. In our case, this is the dfTrain DataFrame. It is a table that contains the numbers that the machine learning algorithm needs to find the pattern.

Please note that we prepared two datasets – dfTrain, which contains the numbers to train the algorithm, and lstSequence, which is the sequence of Fibonacci numbers that we'll find later.

Now, let's start training the algorithm:

```
# algorithm to train
# here, we use linear regression
model = LinearRegression()


# now, the actual process of training the model
model.fit(dfTrain[['first number', 'second number']],
```

```
                              dfTrain['result'])

# printing the score of the model, i.e. how good the model is when
trained
print(model.score(dfTrain[['first number', 'second number']],
dfTrain['result']))
```

The magic of the entire code fragment is in the bold-faced code – the `model.fit` method call. This method trains the logistic regression model based on the data we prepared for it. The model itself is created one line above, in the `model = LinearRegression()` line.

Now, we can make inferences or create new Fibonacci numbers using the following code fragment:

```
# and loop through the newly predicted numbers
for k in range(23):

    # the line below is where the magic happens
    # it takes two numbers from the list
    # formats them to an array
    # and makes the prediction
    # since the model returns a float,
    # we need to convert it to it
    intFibonacci = int(model.predict(np.
array([[lstSequence[k],lstSequence[k+1]]])))

    # add this new number to the list for the next iteration
    lstSequence.append(intFibonacci)

    # and print it
    print(intFibonacci)
```

This code fragment contains a similar line to the previous one – `model.predict()`. This line uses the previously created model to make an inference. Since the Fibonacci sequence is recursive, we need to add the newly created number to the list before we can make the new inference, which is done in the `lstSequence.append()` line.

Now, it is very important to emphasize the difference between these two ways of solving the same problem. The traditional implementation *exposes* the algorithm used to create the numbers. We do not see the Fibonacci sequence there, but we can see how it is calculated. The machine learning implementation *exposes* the data used to create the numbers. We see the first sequence as training data, but we never see how the model creates that sequence. We do not know whether that model is always correct – we would need to test it against the real sequence – simply because we do not know how the algorithm works. This takes us to the next part, which is about just that – probabilities.

Probability and software – how well they go together

The fundamental characteristic that makes machine learning software different from traditional software is the fact that the core of machine learning models is statistics. This statistical learning means that the output of the machine learning model is a probability and, as such, it is not as clear as in traditional software systems.

The probability, which is the result of the model, means that the answer we receive is a probability of something. For example, if we classify an image to check whether it contains a dog or a cat, the result of this classification is a probability – for example, there is a 93% probability that the image contains a dog and a 7% probability that it contains a cat. This is illustrated in *Figure 1.3*:

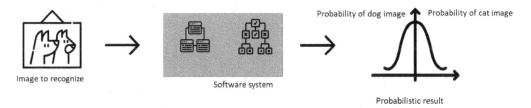

Figure 1.3 – Probabilistic nature of machine learning software

To use these probabilistic results in other parts of the software, or other systems, the machine learning software usually uses thresholds (for example, if $x<0.5$) to provide only one result. Such thresholds specify which probability is acceptable to be able to consider the results to belong to a specific class. For our example of image classification, this probability would be 50% – if the probability of identifying a dog in the image is larger than 50%, then the model states that the image contains a dog (without the probability).

Changing these probabilistic results to digital ones, as we did in the previous example, is often correct, but not always. Especially in corner cases, such as when the probability is close to the threshold's lower bound, the classification can lead to errors and thus to software failures. Such failures are often negligible, but not always. In safety-critical systems, there should be no mistakes as they can lead to unnecessary hazards with potentially catastrophic consequences.

In contexts where the probabilistic nature of machine learning software is problematic, but we still need machine learning for its other benefits, we can construct mechanisms that mitigate the consequences of mispredictions, misclassifications, and sub-optimizations. These mechanisms can guard the machine learning models and prevent them from suggesting wrong recommendations. For example, when we use machine learning image classification in the safety system of a car, we construct a so-called *safety cage* around the model. This safety cage is a non-machine learning component that uses rules to check whether a specific recommendation, classification, or prediction is plausible in the specific context. It can, for instance, prevent a car from suddenly stopping for a non-existent traffic light signal on a highway, which is a consequence of a misclassification of a camera feed from the front camera.

Therefore, let's look at another best practice that encourages the use of machine learning software even in safety-critical systems.

> **Best practice #3**
>
> If your software is safety-critical, make sure that you can design mechanisms to prevent hazards caused by the probabilistic nature of machine learning.

Although this best practice is formulated toward safety-critical systems, it is more general than that. Even for mission-critical or business-critical systems, we can construct mechanisms that can gatekeep the machine learning models and prevent erroneous behavior of the entire software system. An example of how such a cage can be constructed is shown in *Figure 1.4*, where the gatekeeper component provides an additional signal that the model's prediction cannot be trusted/used:

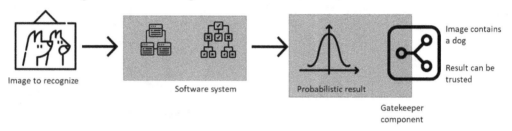

Figure 1.4 – Gatekeeping of machine learning models

In this figure, the additional component is placed as the last one in this processing pipeline to ensure that the result is always binary (for this case). In other cases, such a gatekeeper can be placed in parallel to the machine learning model and can act as a parallel processing flow, where data quality is checked rather than the classification model.

Such gatekeeper models are used quite frequently, such as when detecting objects in perception systems – the model detects objects in individual images, while the gatekeeper checks that the same object is identified consistently over sequences of consecutive images. They can form redundant processing channels and pipelines. They can form feasibility-checking components, or they can correct out-of-bounds results into proper values. Finally, they can also disconnect machine learning components from the pipeline and adapt these pipelines to other components of the software, usually algorithms that make decisions – thus forming self-adaptive or self-healing software systems.

This probabilistic nature of machine learning software means that pre-deployment activities are different from the traditional software. In particular, the process of testing machine learning and traditional software is different.

Testing and evaluation – the same but different

Every machine learning model needs to be validated, which means that the model needs to be able to provide correct inferences for a dataset that the model did not see before. The goal is to assess whether the model has learned patterns in the data, the data itself, or neither. The typical measures of correctness in classification problems are accuracy (the quotient of correctly inferred instances to all classified instances), **Area Under Curve/Receiver Operation Characteristics (AUROC)**, and the **true positive ratio (TPR)** and **false positive ratio (FPR)**.

For prediction problems, the quality of the model is measured in the mispredictions, such as the **mean squared error (MSE)**. These measures quantify the errors in predictions – the smaller the values, the better the model. *Figure 1.5* shows the process for the most common form of supervised learning:

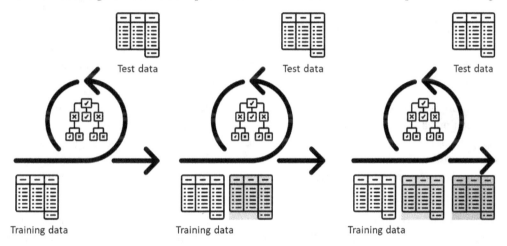

Figure 1.5 – Model evaluation process for supervised learning

In this process, the model is subjected to different data for every iteration of training, after which it is used to make inferences (classifications or regression) on the same test data. The test data is set aside before training, and it is used as input to the model only when validating, never during training.

Finally, some models are reinforcement learning models, where the quality is assessed by the ability of the model to optimize the output according to a predefined function (reward function). These measures allow the algorithm to optimize its operations and find the optimal solution – for example, in genetic algorithms, self-driving cars, or energy grid operations. The challenge with these models is that there is no single metric that can measure performance – it depends on the scenario, the function, and the amount of training that the model received. One famous example of such training is the algorithm from the *War Games* movie (from 1983), where the main supercomputer plays millions of tic-tac-toe games to understand that there is no strategy to win – the game has no winner.

Figure 1.6 presents the process of training a reinforcement system graphically:

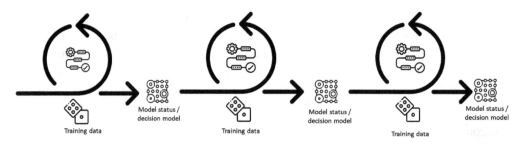

Figure 1.6 – Reinforcement learning training process

We could get the impression that training, testing, and validating machine learning models are all we need when developing machine learning software. This is far from being true. The models are parts of larger systems, which means that they need to be integrated with other components; these components are not validated in the process of validation described in *Figure 1.5* and *Figure 1.6*.

Every software system needs to undergo rigorous testing before it can be released. The goal of this testing is to find and remove as many defects as possible so that the user of the software experiences the best possible quality. Typically, the process of testing software is a process that comprises multiple phases. The process of testing follows the process of software development and aligns with that. In the beginning, software engineers (or testers) use unit tests to verify the correctness of their components.

Figure 1.7 presents how these three types of testing are related to one another. In unit testing, the focus is on algorithms. Often, this means that the software engineers must test individual functions and modules. Integration testing focuses on the connections between modules and how they can conduct tasks together. Finally, system testing and acceptance testing focus on the entire software product. The testers imitate real users to check that the software fulfills the requirements of the users:

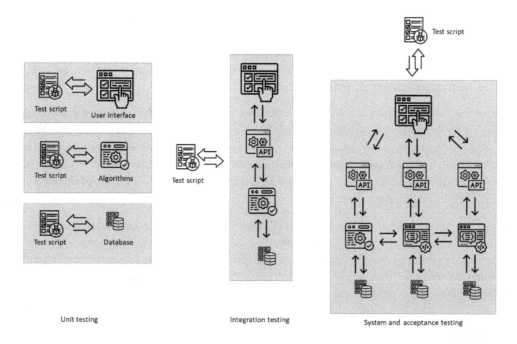

Figure 1.7 – Three types of software testing – unit testing (left), integration testing (middle), and system and acceptance testing (right)

The software testing process is very different than the process of model validation. Although we could mistake unit testing for model validation, this is not entirely the case. The output from the model validation process is one of the metrics (for example, accuracy), whereas the output from the unit test is true/false – whether the software produces the expected output or not. No known defects (equivalent to the false test results) are acceptable for a software company.

In traditional software testing, software engineers prepare a set of test cases to check whether their software works according to the specification. In machine learning software, the process of testing is based on setting aside part of the dataset (the test set) and checking how well the trained model (on the train set) works on that data.

Therefore, here is my fourth best practice for testing machine learning systems.

> **Best practice #4**
>
> Test the machine learning software as an addition to the typical train-validation-evaluation process of machine learning model development.

Testing the entire system is very important as the entire software system contains mechanisms to cope with the probabilistic nature of machine learning components. One such mechanism is the safety cage mechanism, where we can monitor the behavior of the machine learning components and prevent them from providing low-quality signals to the rest of the system (in the case of corner cases, close to the decision boundaries, in the inference process).

When we test the software, we also learn about the limitations of the machine learning components and our ability to handle the corner cases. Such knowledge is important for deploying the system when we need to specify the operational environment for the software. We need to understand the limitations related to the requirements and the specification of the software – the use cases for our software. Even more importantly, we need to understand the implications of the use of the software in terms of ethics and trustworthiness.

We'll discuss ethics in *Chapter 15* and *Chapter 16*, but it is important to understand that we need to consider ethics from the very beginning. If we don't, we risk that our system makes potentially harmful mistakes, such as the ones made by large artificial intelligence hiring systems, face recognition systems, or self-driving vehicles. These harmful mistakes entail monetary costs, but more importantly, they entail loss of trust in the product and even missed opportunities.

Summary

Machine learning and traditional software are often perceived as two alternatives. However, they are more like siblings – one cannot function without the other. Machine learning models are very good at solving constrained problems, but they require traditional software for data collection, preparation, and presentation.

The probabilistic nature of machine learning models requires additional elements to make them useful in the context of complete software products. Therefore, we need to embrace this nature and use it to our advantage. Even for safety-critical systems, we could (and should) use machine learning when we know how to design safety mechanisms to prevent hazardous consequences.

In this chapter, we explored the differences between machine learning software and traditional software while focusing on how to design software that can contain both parts. We also showed that there is much more to machine learning software than just training, testing, and evaluating the model – we showed that rigorous testing makes sense and is necessary for deploying reliable software.

Now, it is time to move on to the next chapter, where we'll open up the black box of machine learning software and explore what we need to develop a complete machine learning software product – starting from data acquisition and ending with user interaction.

References

- Shortliffe, E.H., et al., *Computer-based consultations in clinical therapeutics: explanation and rule acquisition capabilities of the MYCIN system. Computers and biomedical research*, 1975. 8(4): p. 303-320.

- James, G., et al., *An introduction to statistical learning. Vol. 112.* 2013: Springer.

- Saleh, H., *Machine Learning Fundamentals: Use Python and scikit-learn to get up and running with the hottest developments in machine learning.* 2018: Packt Publishing Ltd.

- Raschka, S. and V. Mirjalili, *Python machine learning: Machine learning and deep learning with Python, scikit-learn, and TensorFlow 2.* 2019: Packt Publishing Ltd.

- Sommerville, I., *Software engineering. 10th. Book Software Engineering. 10th, Series Software Engineering*, 2015.

- Houpis, C.H., G.B. Lamont, and B. Lamont, *Digital control systems: theory, hardware, software.* 1985: McGraw-Hill New York.

- Sawhney, R., *Can artificial intelligence make software development more productive? LSE Business Review*, 2021.

- He, X., K. Zhao, and X. Chu, *AutoML: A survey of the state-of-the-art. Knowledge-Based Systems.* 2021. 212: p. 106622.

- Reed, S., et al., *A generalist agent. arXiv preprint arXiv:2205.06175*, 2022.

- Floridi, L. and M. Chiriatti, *GPT-3: Its nature, scope, limits, and consequences. Minds and Machines*, 2020. 30(4): p. 681-694.

- Creswell, A., et al., *Generative adversarial networks: An overview. IEEE signal processing magazine*, 2018. 35(1): p. 53-65.

- Celebi, M.E. and K. Aydin, *Unsupervised learning algorithms.* 2016: Springer.

- Chen, J.X., *The evolution of computing: AlphaGo. Computing in Science & Engineering*, 2016. 18(4): p. 4-7.

- Ko, J.-S., J.-H. Huh, and J.-C. Kim, *Improvement of energy efficiency and control performance of cooling system fan applied to Industry 4.0 data center. Electronics*, 2019. 8(5): p. 582.

- Dastin, J., *Amazon scraps secret AI recruiting tool that showed bias against women. In Ethics of Data and Analytics.* 2018, Auerbach Publications. p. 296-299.

- Castelvecchi, D., *Is facial recognition too biased to be let loose? Nature*, 2020. 587(7834): p. 347-350.

- Siddiqui, F., R. Lerman, and J.B. Merrill, *Teslas running Autopilot involved in 273 crashes reported since last year. In The Washington Post.* 2022.

2
Elements of a Machine Learning System

Data and algorithms are crucial for machine learning systems, but they are far from sufficient. Algorithms are the smallest part of a production machine learning system. Machine learning systems also require data, infrastructure, monitoring, and storage to function efficiently. For a large-scale machine learning system, we need to ensure that we can include a good user interface or package model in microservices.

In modern software systems, combining all necessary elements requires different professional competencies – including machine learning/data science engineering expertise, database engineering, software engineering, and finally interaction design. In these professional systems, it is more important to provide reliable results that bring value to users rather than include a lot of unnecessary functionality. It is also important to orchestrate all elements of machine learning together (data, algorithms, storage, configuration, and infrastructure) rather than optimize each one of them separately – all to provide the most optimal system for one or more use cases from end users.

In this chapter, we'll review each element of a professional machine learning system. We'll start by understanding which elements are important and why. Then, we'll explore how to create such elements and how to put them together into a single machine learning system – a so-called machine learning pipeline.

In this chapter, we're going to cover the following main topics:

- Machine learning is more than just algorithms and data
- Data and algorithms
- Configuration and monitoring
- Infrastructure and resource management
- Machine learning pipelines

Elements of a production machine learning system

Modern machine learning algorithms are very capable because they use large quantities of data and consist of a large number of trainable parameters. The largest available models are **Generative Pre-trained Transformer-3 (GPT-3)** from OpenAI (with 175 billion parameters) and Megatron-Turing from NVidia (356 billion parameters). These models can create texts (novels) and make conversations but also write program code, create user interfaces, or write requirements.

Now, such large models cannot be used on a desktop computer, laptop, or even in a dedicated server. They need advanced computing infrastructure, which can withstand long-term training and evaluation of such large models. Such infrastructure also needs to provide means to automatically provide these models with data, monitor the training process, and, finally, provide the possibility for the users to access the models to make inferences. One of the modern ways of providing such infrastructure is the concept of **Machine learning as a service (MLaaS)**. MLaaS provides an easy way to use machine learning from the perspective of data analysts of software integrators since it delegates the management, monitoring, and configuration of the infrastructure to specialized companies.

Figure 2.1 shows elements of modern machine learning-based software systems. Google has used these to describe production machine learning systems since then. Although variations in this setup exist, the principles remain:

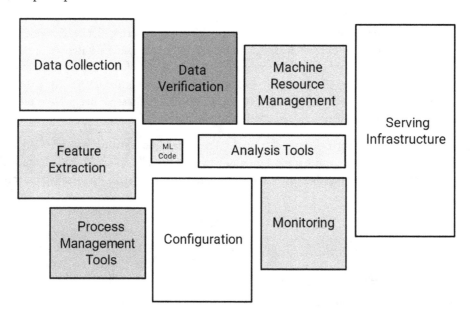

Figure 2.1 – Elements of a production machine learning system

Here, the machine learning model (**ML code**) is the smallest of these elements (Google, under the Creative Commons 4.0 Attribution License, https://developers.google.com/machine-learning/crash-course/production-ml-systems). In terms of the actual source code, in Python, model creation, training, and validation are just three lines of code (at least for some of the models):

```
model = RandomForestRegressor(n_estimators=10, max_depth=2)
model.fit(X_train, Y_train)
Y_pred = model.predict(X_test)
```

The first line creates the model from a template – in this case, it is a random forest model with 10 trees, each of which has a maximum of two splits. Random forest is an ensemble learning method that constructs multiple decision trees during training and outputs the mode of the classes (classification) of the individual trees for a given input. By aggregating the results of multiple trees, it reduces overfitting and provides higher accuracy compared to a single decision tree. The second line trains the model based on the training data (X_train, which contains only the preditors/input features, and Y_train, which contains the predicted class/output features). Finally, the last line makes predictions for the test data (X_test) to compare it to the oracle (the expected value) in subsequent steps. Even though this model.predict(X_test) line is not part of a production system, we still need to make inferences, so there is always a similar line in our software.

Therefore, we can introduce the next best practice.

> **Best practice #5**
>
> When designing machine learning software, prioritize your data and the problem to solve over the algorithm.

In this example, we saw that the machine learning code, from the perspective of the software engineer, is rather small. Before applying algorithms, we need to prepare the data correctly as the algorithms (model.fit(X_train, Y_train)) require the data to be in a specific format – the first parameter is the data that's used to make inferences (so-called feature vectors or input data samples), while the second parameter is the target values (so-called decision classes, reference values, or target values, depending on the algorithm).

Data and algorithms

Now, if using the algorithms is not the main part of the machine learning code, then something else must be – that is, data handling. Managing data in machine learning software, as shown in *Figure 2.1*, consists of three areas:

1. Data collection.
2. Feature extraction.
3. Data validation.

Although we will go back to these areas throughout this book, let's explore what they contain. *Figure 2.2* shows the processing pipeline for these areas:

Raw data source   Raw data   Clean data   Features   Validated data

Figure 2.2 – Data collection and preparation pipeline

Note that the process of preparing the data for the algorithms can become quite complex. First, we need to extract data from its source, which is usually a database. It can be a database of measurements, images, texts, or any other raw data. Once we've exported/extracted the data we need, we must store it in a raw data format. This can be in the form of a table, as shown in the preceding figure, or it can be in a set of raw files, such as images.

Data collection

Data collection is a procedure of transforming data from its raw format to a format that a machine learning algorithm can take as input. Depending on the data and the algorithm, this process can take different forms, as illustrated in *Figure 2.3*:

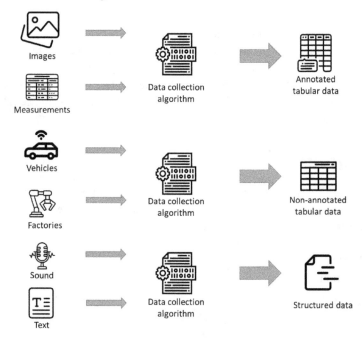

Figure 2.3 – Different forms of data collection – examples

Data from images and measurements such as time series is usually collected to make classifications and predictions. These two classes of problems require the ground truth to be available, which we saw as Y_train in the previous code example. These target labels are either extracted automatically from the raw data or added manually through the process of labeling. The manual process is time-consuming, so the automated one is preferred.

The data that's used in non-supervised learning and reinforcement learning models is often extracted as tabular data without labels. This data is used in the decision process or the optimization process to find the best solution to the given problem. Without optimization, there is a risk that our results are not representative of new data. The preceding figure shows two examples of such problems – optimizations of smart factories of Industry 4.0 and self-driving vehicles. In smart factories, reinforcement learning models are used to optimize production processes or control so-called *dark factories*, which operate entirely without human intervention (the name *dark factories* comes from the fact that there is no need for lights in these factories; robots do not need light to work).

The data that's used for modern self-supervised models often comes from such sources as text or speech. These models do not require a tabular form of the data, but they require structure. For example, to train text transformer models, we need to tokenize the text per sentence (or per paragraph) for the model to learn the context of the words.

Hence, here comes my next best practice.

> **Best practice #6**
> Once you've explored the problem you wish to solve and understood the data availability, decide whether you want to use supervised, self-supervised, unsupervised, or reinforcement learning algorithms.

The fact that we need different data for different algorithms is natural. However, we have not discussed how to decide upon the algorithm. Choosing supervised learning only makes sense when we want to predict or classify data statically – that is, we train the model and then we use it to make inferences. When we train and then make inferences, the model does not change. There is no adjustment as we go and re-training the algorithm is done periodically – I call it a *train once, predict many* principle.

We can choose unsupervised methods when we use/train and apply the algorithm without the target class. Some of these algorithms are also used to group data based on the data's property, for example, to cluster it. I call this the *train once, predict once* principle.

For self-supervised models, the situation is a bit more interesting. There, we can use something called *pre-training*. Pre-training means that we can train a model on a large corpus of data without any specific context – for example, we train language models on large corpora of English texts from Wikipedia. Then, when we want to use the model for a specific task, such as to write new text, we train it a bit more on that task. I call this the *train many, predict once* principle as we must pre-train and train the model for each task.

Finally, reinforcement learning needs data that is changed every time the model is used. For example, when we use a reinforcement learning algorithm to optimize a process, it updates the model each time it is used – we could say it learns from its mistakes. I call this the *train many, predict many* principle.

Usually, the raw data is not ready to be used with machine learning as it can contain empty data points, noise, or broken files. Therefore, we need to clean up these erroneous data points, such as by removing empty data points (using Python commands such as `dataFrame.dropna(...)`) or using data imputation techniques.

Now, there is a fundamental difference between the removal of data points and their imputation. The data imputation process is when we add missing properties of data based on similar data points. It's like filling in blanks in a sequence of numbers – 1, 2, 3, …, 5, where we fill in the number 4. Although filling in the data increases the number of data points available (thus making the models better), it can strengthen certain properties of the data, which can cause the model to learn. Imputation is also relevant when the size of the data is small; in large datasets, it is better (more resource-efficient and fair) to drop the missing data points. With that, we've come to my next best practice.

> **Best practice #7**
>
> Use data imputation only when you know which properties of data you wish to strengthen and only do so for small datasets.

Finally, once we have clean data to work with, we can extract features. There, we can use algorithms that are specific to our problem at hand. For example, when we work with textual data, we could use a simple bag-of-words to count the frequencies of words, though we can also use the word2vec algorithm to embed the frequencies of the co-occurrence of words (algorithms that we'll discuss in the next few chapters). Once we've extracted features, we can start validating the data. The features can emphasize certain properties of data that we didn't see before.

One such example is noise – when we have data in a feature format, we can check whether there is *attribute* or *class noise* in the data. Class noise is a phenomenon that is related to labeling errors – one or more data points have been labeled incorrectly. Class noise can be either contradictory examples or wrongly labeled data points. It is a dangerous phenomenon since it can cause low performance when training and using machine learning models.

Attribute noise is when one (or more) attributes is corrupted with wrong values. Examples include wrong values, missing attribute (feature) values, and irrelevant values.

Once the data has been validated, it can be used in algorithms. So, let's dive a bit deeper into each step of the data processing pipeline.

Now, since different algorithms use data in different ways, the data has a different form. Let's explore how the data should be structured for each of the algorithms.

Feature extraction

The process of transforming raw data into a format that can be used by algorithms is called feature extraction. It is a process where we apply a feature extraction algorithm to find properties of interest in the data. The algorithm for extracting features varies depending on the problem and the data type.

When we work with textual data, we can use several algorithms, but let me illustrate the use of one of the simplest ones – bag-of-words. The algorithm simply counts the occurrence of words in the sentence – it either counts a pre-defined set of words or uses statistics to find the most frequent words. Let's consider the following sentence:

```
Mike is a tall boy.
```

When we use the bag-of-words algorithm without any constraints, it provides us with the following table:

Sentence ID	Mike	Is	A	tall	Boy
0	1	1	1	1	1

Figure 2.4 – Features extracted using bag-of-words

The table contains all the words in the sentence as features. It is not very useful for just one sentence, but if we add another one (sentence 1), things become more obvious. So, let's add the following sentence:

```
Mary is a smart girl.
```

This will result in the following feature table:

Sentence ID	Mike	Is	A	Tall	boy	smart	girl
0	1	1	1	1	1	0	0
1	0	1	1	0	0	1	1

Figure 2.5 – Features extracted from two sentences

We are now ready to add the label column to the data. Let's say that we want to label each sentence as being positive or negative. The table then gets one more column – `label` – where **1** means that the sentence is positive and **0** otherwise:

Sentence ID	Mike	Is	A	Tall	boy	smart	girl	Label
0	1	1	1	1	1	0	0	1
1	0	1	1	0	0	1	1	1

Figure 2.6 – Labels added to the data

Now, these features allow us to see the difference between two sentences, which we can then use to train and test machine learning algorithms.

There are, however, two important limitations of this approach. The first one is that it is impractical (if not impossible) to have all words from all sentences as columns/features. For any non-trivial text, this would result in large and sparse matrices – a lot of wasted space. The second limitation is the fact that we usually lose important information – for example, the sentence "Is Mike a boy?" would result in the same feature vector as the first sentence. A feature vector is an n-dimensional vector of numerical features that describe some object in pattern recognition in machine learning. Although these sentences are not identical, they become undistinguishable, which can lead to class noise if they are labeled differently.

The problem with adding this kind of noise becomes even more evident if we use statistics to choose the most frequent words as features. Here, we can lose important features that discriminate data points in a useful way. Therefore, this bag-of-words approach is only for illustration here. Later in this book, we'll dive deeper into so-called transformers, which use more advanced techniques to extract features.

Data validation

Feature vectors are the core of machine learning algorithms. They are most prominently, and directly, used by supervised machine learning algorithms. However, the same concepts of data validation apply to data used in other types of validation.

Every form of data validation is a set of checks that ensure the data contains the desired properties. An example of such a set of checks is presented in *Figure 2.7*:

Figure 2.7 – An example of data quality checks

Completeness of data is a property that describes how much of the total distribution our data covers. This can be measured in terms of object distribution – for example, how many types/models/colors of cars we have in our image dataset – or it can be measured in terms of properties – for example, how many of the words in the language our data contains.

Accuracy is a property that describes how well our data is related to the empirical (real) world. For example, we may want to check whether all images in our dataset are linked to an object or whether all objects in the image are annotated.

Consistency describes how well the data is structured internally and whether the same data points are annotated in the same way. For example, in binary classification, we want all data points to be labeled "0" and "1" or "True" and "False," but not both.

Integrity is the property where we check that the data can be integrated with other data. The integration can be done through common keys or other means. For example, we can check if our images contain metadata that will allow us to know where the image was taken.

Finally, timeliness is a property that describes how fresh the data is. It checks whether the data contains the latest required information. For example, when we design a recommendation system, we would like to recommend both the new items and the old ones, so timeliness is important.

So, here is the next best practice.

> **Best practice #8**
> Choose the data validation attributes that are the most relevant for your system.

Since every check requires additional steps in the workflow and can slow down the processing of the data, we should choose the data quality checks that impact our business and our architecture. If we develop a system where we want to provide up-to-date recommendations, then timeliness is our top priority, and we should focus on that rather than on completeness.

Although there are a lot of frameworks for data validation and assessing data quality, I usually use a subset of data quality attributes from the AIMQ framework. The AIMQ framework has been designed to quantify data quality based on several quality attributes, similar to quality frameworks in software engineering such as the ISO 25000 series. I find the following properties of data to be the most important to validate:

- The data is free of noise
- The data is fresh
- The data is fit for purpose

The first property is the most important as noisy data can cause low performance in machine learning algorithms. For class noise, which we introduced previously, it is important to check if the data labels are not contradictory and to check whether the labels are assigned correctly. Contradictory labels can be found automatically, but wrongly annotated data points need manual assessment. For attribute noise, we can use statistical approaches to identify attributes that have low variability (or even the ones that are constant) or attributes that are completely random and do not contribute to the model's learning. Let's consider an example of a sentence:

```
Mike is not a tall boy.
```

If we use the same feature extraction technique as for the previous sentences, our feature matrix looks like what's shown in *Figure 2.8*. We use the same features as for sentences 0 and 1 for sentence 2. Since the last sentence differs only in terms of one word (not), this can lead to class noise. The last column has a label, **0**, which indicates that the sentence is negative. However, since we use the same feature extraction algorithm, the feature vector does not include the word not. This can happen when we train the model on one dataset and apply it to another one. This means that the first sentence and the last sentence have identical feature vectors, but different labels:

Sentence ID	Mike	Is	A	tall	boy	smart	girl	Label
0	1	1	1	1	1	0	0	1
1	0	1	1	0	0	1	1	1
2	1	1	1	1	1	0	0	0

Figure 2.8 – Noisy dataset

Having two different annotations for the same feature vectors is problematic as the machine learning algorithms cannot learn the pattern since there isn't one for these noisy data points. Therefore, we need to validate the data in terms of it being free from noise.

Another property that the data needs to possess is its timeliness – that is, being fresh. We must use current, not old, data. One of the areas where this is of utmost importance is autonomous driving, where we need to keep the models up to date with the latest conditions (for example, the latest traffic signs).

Finally, the most important part of validation is assessing whether the data is fit for purpose. Note that this assessment cannot be done automatically as it needs to be done expertly.

Configuration and monitoring

Machine learning software is meant to be professionally engineered, deployed, and maintained. Modern companies call this process *MLOps*, which means that the same team needs to take responsibility for both the development and operations of the machine learning system. The rationale behind this extended responsibility is that the team knows the system best and therefore can configure, monitor, and maintain it in the best possible way. The teams know the design decisions that must be taken when developing the system, assumptions made about the data, and potential risks to monitor after the deployment.

Configuration

Configuration is one such design decision that's made by the development team. The team configures the parameters of the machine learning models, the execution environment, and the monitoring infrastructure. Let's explore the first one; the latter two will be discussed in the next few sections.

To exemplify this challenge, let's look at a random forest classifier for a dataset for classifying events during a specific surgery. The classifier, at least in its Python implementation, has 16 hyperparameters (`https://scikit-learn.org/stable/modules/generated/sklearn.ensemble.RandomForestClassifier.html`). Each of these hyperparameters has several values, which means that finding the optimal set of hyperparameters can be a tedious task.

However, in practice, we do not need to explore all hyperparameters' values and we do not need to explore all combinations. We should only explore the ones that are the most relevant for our task and, by extension, the dataset. I usually explore only two hyperparameters as these are the most important: the number of trees and the depth of the tree. The first determines how broad the forest is, while the second determines how deep it is. The code to specify these parameters can look something like this:

```
rnd_clf = RandomForestClassifier(n_estimators=2,
                                 max_leaf_nodes=10,
                                 n_jobs=-1)
```

The `n_estimators` hyperparameter is the number of trees, while `max_depth` hyperparameter is the depth of each tree. The values of these parameters depend on the dataset – how many features we have and how many data points we have. If we have too many trees and too many leaves compared to the number of features and data points, the classifier gets overfitted and cannot generalize from the data. This means that the classifier has learned to recognize each instance rather than recognize patterns in the data – we call this *overfitting*.

If we have too few trees or leaves, then the generalized patterns will be too broad and therefore we observe errors in classification – at least more errors than the optimal classifier. We call this *underfitting* as the model does not learn the pattern correctly.

So, we can write a piece of code that would search for the best combination of these two parameters based on the pre-defined set of values. The code to find the best parameters manually would look something like this:

```
numEstimators = [2, 4, 8, 16, 32, 64, 128, 256, 512]
numLeaves = [2, 4, 8, 16, 32, 64, 128]
for nEst in numEstimators:
  for nLeaves in numLeaves:
    rnd_clf = RandomForestClassifier(n_estimators=nEst,
                                     max_leaf_nodes=nLeaves,
                                     n_jobs=-1)
    rnd_clf.fit(X_train, y_train)
```

```
y_pred_rf = rnd_clf.predict(X_test)
accuracy_rf = accuracy_score(y_test, y_pred_rf)
print(f'Trees: {nEst}, Leaves: {nLeaves}, Acc: {accuracy_rf:.2f}')
```

The two lines emphasized in orange show two loops that explore these parameters – the content of the inner loop trains the classifier with these parameters and prints out the output.

Let's apply this algorithm to physiological data from patients who have undergone operations. When we plot the output on the diagram, as shown in *Figure 2.9*, we can observe how the accuracy evolves. If we set the number of trees to 2, the classifier's best performance is the best for 8 leaves, but even then, it does not classify the events perfectly. For four trees, the classifier achieves the best performance with 128 leaves, and the accuracy is 1.0. From the following diagram, we can see that adding more trees does not improve the results significantly:

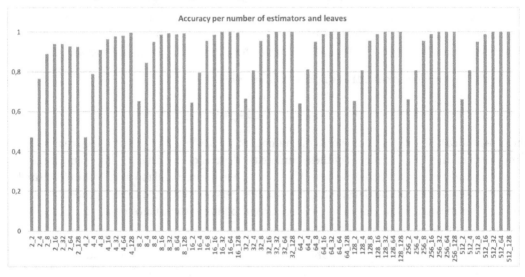

Figure 2.9 – Accuracy per number of estimators and leaves. The labels of the x axis show the number of trees (before the underscore) and the number of leaves (after the underscore)

For this example, the time required to search for the best result is relatively short – it takes up to 1-2 minutes on a standard laptop. However, if we want to find the optimal combination of all 16 parameters, we will spend a significant amount of time doing this.

There is a more automated way of finding the optimal parameters of machine learning classifiers – different types of search algorithms. One of the most popular ones is the GridSearch algorithm (https://scikit-learn.org/stable/modules/generated/sklearn.model_selection.GridSearchCV.html), which works similarly to our manual script, except that it can do cross-validation with multiple splits and many other statistical tricks to improve the search. Searching for the optimal solution with the GridSearch algorithm can look something like this:

```
# Create the parameter grid based on the results of random search
param_grid = {
```

```
      'max_depth': [2, 4, 8, 16, 32, 64, 128],
      'n_estimators': [2, 4, 8, 16, 32, 64, 128, 256, 512]
}
# Create a base model
rf = RandomForestClassifier()

# Instantiate the grid search model
grid_search = GridSearchCV(estimator = rf,
                           param_grid = param_grid,
                           cv = 3,
                           n_jobs = -1)

# Fit the grid search to the data
grid_search.fit(X_train, y_train)

# get the best parameters
best_grid = grid_search.best_estimator_

# print the best parameters
print(grid_search.best_params_)
```

The preceding code finds the best solution and saves it as the `best_estimator_` parameter of the GridSearch model. In the case of this dataset and this model, the algorithm finds the best random forest to be the one with 128 trees (`n_estimators`) and 4 levels (`max_depth`). The results are a bit different than the ones found manually, but this does not mean that one of the methods is superior.

However, having too many trees can result in overfitting, so I would choose the model with 4 trees and 128 leaves over the one with 128 trees and 4 levels. Or maybe I would also use a model that is somewhere in-between – that is, a model that has the same accuracy but is less prone to overfitting in either the depth or the width.

This leads to my next best practice.

Best practice #9

Use GridSearch and other algorithms after you have explored the parameter search space manually.

Although the automated parameter search algorithms are very useful, they hide properties of the data from us and do not allow us to explore the data and the parameters ourselves. From my experience, understanding the data, the model, its parameters, and its configuration is crucial to the successful deployment of machine learning software. I only use GridSearch (or other optimization algorithms) after I've tried to find some optima manually since I would like to understand the data.

Monitoring

Once the machine learning system has been configured, it is set into production, often as part of the larger software system. The inferences that are made by machine learning are the basis for the features of the product and the business model behind this. Therefore, the machine learning component should make as few mistakes as possible. Unfortunately, the customers take failures and mistakes more seriously than correctly functioning products.

However, the performance of the machine learning system degrades over time, but not because of low-quality programming or design – this is the nature of probabilistic computing. Therefore, all machine learning systems need to be monitored and maintained.

One of the aspects that needs to be monitored is called *concept drift*. Concept drift is a phenomenon in the data, which means that the distribution of entities in the data changes over time for natural reasons. *Figure 2.10* illustrates concept drift for a machine learning classifier (blue and red lines) of images of yellow and orange trucks:

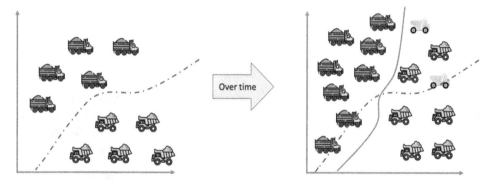

Figure 2.10 – Illustration of concept drift. The original distribution of the objects (on the left) changes over time (on the right), so the classifier must be retrained (blue versus red lines)

The left-hand side shows the original distribution of data that was used to train the model. The model is conceptually shown as the blue dotted line. The model recognizes the differences between the two classes of images. However, over time, the data can change. New images can appear in the dataset and the distribution can change. The original model starts to make mistakes in inference and therefore needs to be adjusted. The re-trained model – that is, the solid red line – captures the new distribution of the data.

It is this change in the dataset that we call concept drift. It is more common in complex datasets and supervised learning models, but the effects of concept drift are equally problematic for non-supervised models and reinforcement learning models.

Figure 2.11 presents the performance of the same random forest model applied to the data from the same distribution (directly after training) and on the data after some operation. Concept drift is visible in the accuracy reduction from 1.0 to 0.44. The model has been trained on the same data as in the example in *Figure 2.9* but has been applied to data from another patient:

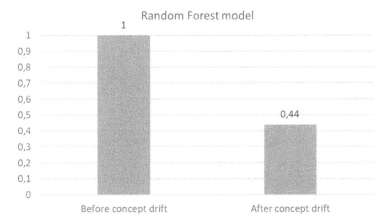

Figure 2.11 – An example of performance decrease before and after concept drift

Therefore, I would like to introduce my next best practice.

Best practice #10

Always include monitoring mechanisms in your machine learning systems.

Including mechanisms to monitor concept drift, even a simple mechanism such as using a Chi-square statistical test for the similarity of distributions, makes a lot of difference. It allows us to identify problems in the system, troubleshoot them, and prevent them from spreading to other parts of the software, or even to the end user of the software.

Professional machine learning engineers set up monitoring mechanisms for concept drift in production, which indicates the degradation of AI software.

Infrastructure and resource management

The infrastructure and resources needed for the machine learning software are organized into two areas – data serving infrastructure (for example, databases) and computational infrastructure (for example, GPU computing platforms). There is also serving infrastructure, which is used to provide the services to the end users. The serving infrastructure could be in the form of desktop applications, embedded software (such as the one in autonomous vehicles), add-ins to tools (as in the case of GitHub Co-pilot), or websites (such as ChatGPT). However, in this book, we'll focus on the data-serving infrastructure and the computational infrastructure.

Both areas can be deployed locally or remotely. Local deployment means that we use our own infrastructure at the company, while remote infrastructure means that we use cloud services or services of another supplier.

Conceptually, we could see these two areas as co-dependent, as depicted in *Figure 2.12*:

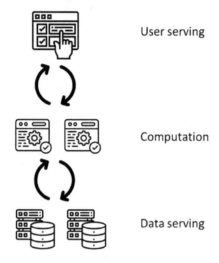

Figure 2.12 – Co-dependency between serving, computing, and data-serving infrastructure

The data serving infrastructure provisions data that's used for the computation infrastructure. It consists of databases and other data sources (for example, raw files). The computation infrastructure consists of computing infrastructure to train and test machine learning models. Finally, the user-serving infrastructure uses the models to make inferences and provides services and functionality to the end users.

Data serving infrastructure

Data serving infrastructure is one of the fundamental parts of the machine learning software because there is no machine learning if there is no data. Data-hungry machine learning applications pose new requirements for the infrastructure in terms of performance, reliability, and traceability. The last requirement is very important as the machine learning training data determines how the trained machine learning model makes inferences. In the case of defects in the end user function, software engineers need to scrutinize the algorithms, the models, and the data that were used to construct the failing machine learning system.

In contrast to traditional software, the data-serving infrastructure is often composed of three different parts, as shown in *Figure 2.13*:

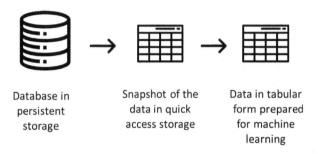

Database in persistent storage

Snapshot of the data in quick access storage

Data in tabular form prepared for machine learning

Figure 2.13 – Data serving infrastructure

The data is stored in persistent storage, such as in a database on a hard drive. It can be stored locally or on a cloud server. The most important part is that this data is secure and can be accessed for further processing. The persistently stored data needs to be extracted so that it can be used in machine learning. The first step is to find the snapshot of the data that is needed – for example, by selecting data for training. The snapshot needs to be prepared and formatted in tabular form so that the machine learning algorithms can use the data to make inferences.

There are several different types of databases today that machine learning systems use. First, there are standard relational databases, where data is stored in the form of tables. These are the databases that are well known and used widely both in traditional and machine learning software.

Newer types of databases are non-SQL databases such as Elasticsearch (`https://www.elastic.co`), which are designed to store documents, not tables. These documents are indexed flexibly so that data can be stored and retrieved based on these documents. In the case of machine learning software, these documents can be images, entire text documents, or even sound data. This is important for storing data in the same format as it is collected so that we can trace the data when needed.

Regardless of the format of the data in the databases, it is retrieved from the database and transformed into tabular form; we'll discuss this in *Chapter 3*. This tabular form is required for the data to be processed by the computational infrastructure.

With that, we've come to my next best practice.

> **Best practice #11**
> Choose the right database for your data – look at this from the perspective of the data, not the system.

Although it sounds obvious that we should choose the right database for the data that we have, for machine learning systems, it is important to select the database that works best for the data at hand, not the system. For example, when we use natural language processing models, we should store the data in documents that we can easily retrieve in an organized form.

Computational infrastructure

The computational infrastructure can change over time. In the early phases of the development of machine learning systems, software developers often use pre-configured experimentation environments. These can be in the form of Jupyter Notebooks on their computers or in the form of pre-configured services such as Google Colab or Microsoft Azure Notebooks. This kind of infrastructure supports rapid prototyping of machine learning, easy provisioning of data, and no need for setting up advanced features. They also allow us to easily scale the computational resources up and down without the need to add or remove extra hardware.

An alternative to this approach is to use our own infrastructure, where we set up your own servers and runtime environments. It requires more effort, but it provides us with full control over the computational resources, as well as full control over data processing. Having full control over data processing can sometimes be the most important factor for selecting the infrastructure.

Hence, my next best practice.

> Best practice #12
>
> Use cloud infrastructure if you can as it saves resources and reduces the need for specialized competence.

Professional AI engineers use self-owned infrastructure for prototyping and training and cloud-based infrastructure for production as it scales better with the number of users. The opposite, which is to use our own infrastructure, is true only if we need to retain full control over data or infrastructure. Full control can be required for applications that use sensitive customer data, military applications, security applications, and other applications where data is extremely sensitive.

Luckily, we have several large actors providing the computational infrastructure, as well as a large ecosystem of small actors. The three largest – Amazon Web Services, Google Cloud, and Microsoft Azure – can provide all kinds of services for both small and large enterprises. Amazon Web Services (`https://aws.amazon.com`) specializes in provisioning data storage and processing infrastructure. It is often used for applications that must quickly process large quantities of data. The infrastructure is professionally maintained and can be used to achieve near-perfect reliability of the products built on that platform. To use it efficiently, you must usually work with containers and virtual machines that execute the code of the machine learning application.

Google Cloud (`https://cloud.google.com`) specializes in provisioning platforms for data-intensive applications and computation-intensive solutions. Thanks to Google's processors (**tensor processing units (TPUs)**), the platform provides a very efficient environment for both training and using machine learning solutions. Google Cloud also provides free solutions for learning machine learning in the form of Google Colab, which is an extension of the Jupyter Notebook (`https://jupyter.org`) platform on Python.

Microsoft Azure (`https://azure.microsoft.com`) specializes in provisioning platforms for training and deploying machine learning systems in the form of virtual machines. It also provides ready-to-deploy models for image recognition, classification, and natural language processing, and even platforms for training generic machine learning models. It is the most flexible of the available platforms and the most scalable.

In addition to these platforms, you can use several specialized ones, such as Facebook's platform for machine learning, which specializes in recommender systems. However, since the specialized platforms are often narrow, we need to remember the potential issues if we want to port our software from one platform to another.

Hence, my next best practice.

> **Best practice #13**
>
> Decide on which production environment you wish to use early and align your process with that environment.

We need to decide if we want to use Amazon's, Google's, or Microsoft's cloud environment or whether we want to use our own infrastructure to reduce the cost of software development. Although it is possible to move our software between these environments, it is not straightforward and requires (at best) significant testing and pre-deployment validation, which often comes with significant costs.

How this all comes together – machine learning pipelines

In this chapter, we explored the main characteristics of machine learning systems and compared them to traditional software systems. Let's finish this comparison by summarizing how we usually design and describe machine learning systems – by using pipelines. A pipeline is a sequence of data processing steps, including the machine learning models. The typical set of steps (also called phases) is shown in *Figure 2.14*:

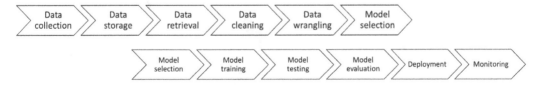

Figure 2.14 – A typical sequence of steps in a machine learning pipeline

This kind of pipeline, although drawn linearly, is usually processed in cycles, where, for example, monitoring for concept drift can trigger re-training, re-testing, and re-deployment.

Machine learning pipelines, just like the one presented in *Figure 2.14*, are often depicted as a set of components as parts of the entire system. However, presenting it using the pipeline analogy helps us understand that machine learning systems process data in steps.

In the next chapter, we'll explore the first part of the pipeline – working with data. We'll start by exploring different types of data and how these types of data are collected, processed, and used in modern software systems.

References

- *Shortliffe, E.H., et al., Computer-based consultations in clinical therapeutics: explanation and rule acquisition capabilities of the MYCIN system. Computers and biomedical research, 1975. 8(4): p. 303-320.*

- *Vaswani, A., et al., Attention is all you need. Advances in neural information processing systems, 2017. 30.*

- *Dale, R., GPT-3: What's it good for? Natural Language Engineering, 2021. 27(1): p. 113-118.*

- *Smith, S., et al., Using deepspeed and megatron to train megatron-turing nlg 530b, a large-scale generative language model. arXiv preprint arXiv:2201.11990, 2022.*

- *Lee, Y.W., et al., AIMQ: a methodology for information quality assessment. Information & management, 2002. 40(2): p. 133-146.*

- *Zenisek, J., F. Holzinger, and M. Affenzeller, Machine learning based concept drift detection for predictive maintenance. Computers & Industrial Engineering, 2019. 137: p. 106031.*

- *Amershi, S., et al. Software engineering for machine learning: A case study. in 2019 IEEE/ACM 41st International Conference on Software Engineering: Software Engineering in Practice (ICSE-SEIP). 2019. IEEE.*

3
Data in Software Systems – Text, Images, Code, and Their Annotations

Machine learning (ML) systems are data-hungry applications, and they like their data well prepared for training and inference. Although it may sound obvious, it is more important to scrutinize the properties of data than to select an algorithm to process the data. The data, however, can come in many different formats and can be from different sources. We can consider data in its raw format – for example, a text document or an image file. We can also consider data in a format that is specific to a task at hand – for example, tokenized text (where words are divided into tokens) or an image with bounding boxes (where objects are identified and enclosed in rectangles).

When considering the end user system, what we can do with the data and how we handle the data becomes crucial. However, identifying important elements in the data and transforming it into a format that is useful for ML algorithms depends on what our task is and which algorithm we use. Therefore, in this chapter, we will work both with data and with algorithms to process it.

In this chapter, we will introduce three data types – images, text, and formatted text (program source code). We will explore how each of these types of data can be used in ML, how they should be annotated, and for what purpose.

Introducing these three types of data provides us with the possibility to explore different ways of annotating these sources of data. Therefore, in this chapter, we will focus on the following:

- Raw data and features – what are the differences?
- Every data has its purpose – annotations and tasks
- Where different types of data can be used together – an outlook on multi-modal data models

Raw data and features – what are the differences?

ML systems are data-hungry. They rely on the data to be trained and to make inferences. However, not all data is equally important. Before the era of **deep learning** (**DL**), the data was supposed to be processed in order to be used in ML. Before DL, the algorithms were limited in the amount of data that could be used for training. The storage and memory limitations were also limited, and therefore, ML engineers had to prepare the data much more than for DL. For example, ML engineers needed to spend more effort to find a small but still representative sample of data for training. After the introduction of DL, ML models can find complex patterns in much larger datasets. Therefore, the work of ML engineers is now focused on finding sufficiently large, and representative, datasets.

Classical ML systems – that is, non-DL systems – require data in a tabular form in order to make inferences, and therefore it is important to design the right feature extraction mechanisms for this kind of system.

DL systems, on the other hand, require minimal data processing and can learn patterns from data in its (almost) raw format. Minimal processing of data is needed as DL systems need a bit of different information about the data for different tasks; they also extract information from the raw data by themselves. For example, they can capture the context of a text without the need to manually process it. *Figure 3.1* illustrates these differences between different types of data based on the tasks that can be performed on them. In this case, the data is in the form of images:

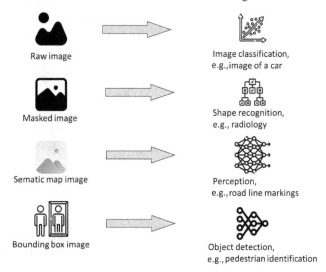

Figure 3.1 – Types of learning systems and the data that they require for images

Raw images are often used for further processing, but they can be used in such tasks as image classification. The task of image classification relates to when the input to the algorithm is the raw image, and the output is the class of the image. We often see these kinds of tasks when we talk about images that contain "cats," "dogs," or "cars."

There are considerable practical applications of this task. One application is in insurance. Several insurance companies have changed their business model and digitalized their businesses. Before the mid-2010s, insurance companies required an initial visit to a workshop to make a first assessment of damage to cars. Today, that first damage is assessed automatically by image classification algorithms. We take a picture of the damaged part with a smartphone and send it to the insurance company's software, where trained ML algorithms are used to make an assessment. In rare, difficult cases, the image needs to be scrutinized by a human operator. This kind of workflow saves money and time and provides a better experience for the handling of damage claims.

Another application is medical image classification, where radiology images are classified automatically to provide an initial diagnosis, and therefore, reduce the burden on medical specialists (in this case, radiologists).

Masked images are processed using filters to emphasize aspects of interest. Most often, these filters are black-and-white filters or grayscale filters. They emphasize the differences between light and dark parts of the images to make it easier to recognize shapes and then to classify these shapes and trace them (in the case of video feeds). This kind of application is often used in perception systems – for example, in cars.

One practical application of a perception system that uses masked images is the recognition of horizontal road markings, such as lane markings. The vehicle's camera takes a picture of the road in front of the car, then its software masks the image and sends it to an ML algorithm for detection and classification. **OpenCV** is one of the libraries used for this kind of task. Other practical applications include face recognition or **optical character recognition** (**OCR**).

Semantic map images include overlays that describe what can be seen in the image, covering a part of an image that contains specific information such as the sky, a car, a person, or a building. A semantic map can cover a part of the image that contains a car, the road that the car is on, the surroundings, and the sky. The semantic map provides rich information about the image that is used for advanced vision perception algorithms, which, in turn, provide information for decision algorithms. Vision perception is particularly needed in automotive systems for autonomous vehicles.

One of the applications of semantic maps is in the active safety systems of vehicles. Images captured by the front camera are processed using **convolutional neural networks** (**CNNs**) to add a semantic map and then used in decision algorithms. These decision algorithms either provide feedback to the driver or take actions autonomously. We can see that often when a car reacts to driving too close to another car or when it detects an obstacle in its way.

Other applications of semantic maps include medical image analyses, whereby ML algorithms provide input to medical specialists as to what the image contains. An example can be brain tumor segmentation using **deep CNNs** (**DCNNs**).

Finally, bounding-box images contain information about the boundaries of objects in images. For each shape of interest, such as a car, pedestrian, or tumor, there is a bounding box surrounding that part of the image, annotated with the class of that shape. These kinds of images are used for detecting objects and providing that information to other algorithms.

One of the applications where we use this kind of image is object recognition in robot coordination systems. A robot's camera registers an image, the CNN identifies objects, and the robot's decision software traces the object in order to avoid collisions. Tracing each object is used to change the behavior of the autonomous robot to reduce the risk of collisions and damage, as well as to optimize the operations of the robot and its environment.

Hence my first best practice in this chapter.

> **Best practice #14**
> Design the entire software system based on the task that you need to solve, not only the ML model.

Since every kind of algorithm that we use requires different processing of the image and provides different kinds of information, we need to understand how to create the entire system around it. In the previous chapter, we discussed pipelines, which include only the ML data pipeline, but a software system requires much more. For safety-critical functionality, we need to design safety cages and signaling to reduce the risk of wrong classifications/detections from ML models. Therefore, we need to understand what we want to do – whether the information is only for making simple decisions (for example, damaged bumper in a car versus not) or whether the classification is part of complex behavior decisions (for example, should the robot turn to the right to avoid the obstacle or should it slow down to let another robot move past?).

Images are one type of data that we use in ML; another one is text. The use of text has been popularized in recent years with the introduction of **recurrent NNs (RNNs)** and transformers. These NN architectures are DL networks that are capable of capturing the context (and, by extension, basic semantics) of words. These models find statistical connections between tokens (and therefore, words) and so can identify similarities that classical ML models are not capable of. Machine translations were a popular application of these models in the beginning, but now, the applications are much wider than this – for example, in understanding programming languages. *Figure 3.2* shows the type of text data that can be used with different types of models:

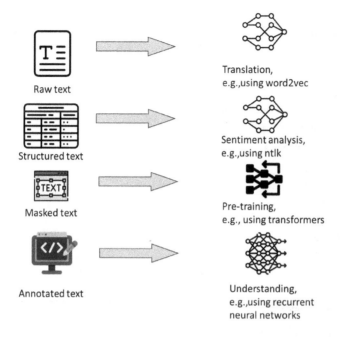

Figure 3.2 – Types of learning systems and the data they require for text

The raw text data is used today in training **large language models** (**LLMs**), but historically, this kind of data was used for various machine translation models. One of the models that is used in these tasks is the word2vec model, which translates text tokens into vectors of numbers – embeddings – which are distances from that token to other tokens in the vocabulary. We saw an example of this in *Chapter 2*, where we counted the number of words in sentences. By using this technology, the word2vec model captures the context of tokens, also known as their similarity. This similarity can be extended to entire sentences or even paragraphs, depending on the size and depth of the model.

Another application of raw text, although in a structured format, is in **sentiment analysis** (**SA**). We use a tabular format of the text data in order to analyze whether the text's sentiment is positive, negative, or neutral. An extension of that task is understanding the intent of the text – whether it is an explanation, a query, or a description.

Masked text data refers to when we mask one or more tokens in a sequence of tokens, and we train the models to predict the token. This is an example of self-supervised training as the model is trained on data that is not annotated, but by masking tokens in different ways (for example, random, based on similarity, human annotations), the model can understand which tokens can be used in that specific context. The larger the model – the transformer – the more data is needed, and a more complex training process is needed.

Finally, annotated text refers to when we label pieces of text with a specific class, just as with images. An example of such annotation is a sentiment. Then, the model captures patterns in the data, and therefore, can repeat these patterns. An example of a task in this area is sentiment recognition, where the model is trained to recognize whether a piece of text is positive or negative in its tone.

A special case of textual data is programming language source code. The use of ML models for programming language tasks has become more popular in the last few years as it provides the possibility to increase the speed and quality of software development. *Figure 3.3* shows types of programming language data and typical tasks:

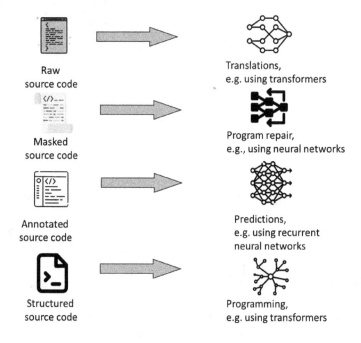

Figure 3.3 – Types of programming language data and typical tasks

Raw source code data is used for tasks that are related to programming language understanding – for example, translations between different programming languages, such as the one using the TransCoder model. This task is similar to translation between natural languages, although it adds additional steps to make the program compile and pass test cases.

Masked programming language code is often used for training models with the purpose of repairing defects – the model is trained on a set of programs that correct defects and then applied on programs with defects. Masked programs are used to train models that can identify problems and provide fixes for them. These tasks are quite experimental at the moment of writing this book but with very promising results.

Annotated source code is used for a variety of tasks. These tasks include defect predictions, code reviews, and identification of design patterns or company-specific design rules. ML models provide much better results than any other techniques for these tasks – for example, compared to static code analysis tools.

Source code is used to train models for advanced software engineering tasks, such as creating programs. GitHub Copilot is one such tool that has been very successful, both in research and in commercial applications.

Now, the aforementioned three types of data illustrate only a small number of applications of ML. The sky is the limit for those who want to utilize ML models for designing software systems. Before designing the systems, however, we need to understand how we work with data in more detail.

Images

Raw image data is often stored in files with annotations in other files. Raw image data presents aspects that are relevant to the system in question. An example of data used for training active safety algorithms is presented in *Figure 3.4*:

Figure 3.4 – Front-camera image from a vehicle

The image with the car is used, in this example, to train a CNN to recognize whether it is safe to drive (for example, whether the road ahead is free from obstacles). With the data annotated on the image level – that is, without masks and bounding boxes – the ML models can either classify the entire image or identify objects. When identifying objects, the models add bounding-box information to the images.

In order to train a CNN for images that contain many objects of significant size (such as high-definition resolution of 1920 x 1080 pixels), we need both large datasets and large computational resources. There are a few reasons for this.

First, colors require a lot of data to be recognized correctly. Although we humans see the color red as almost uniform, the actual pixel intensity of that color varies a lot, which means that we need to create a CNN so that it can understand that different shades of red are sometimes important to recognize braking vehicles.

Second, the large size of images contains details that are not relevant. *Figure 3.5* presents how a CNN is designed. This is a CNN in a LeNet style:

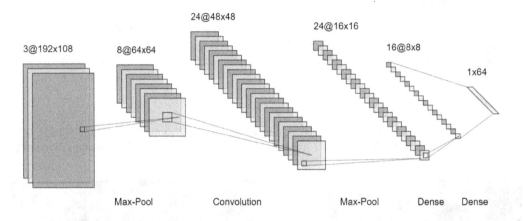

Figure 3.5 – Conceptual design of a CNN

Figure 3.5 shows that the NN takes as input an image of size 192 x 108 pixels (10 times smaller than an HD-quality image). It then uses `MaxPool` layers (for example) to reduce the number of elements, and then it uses convolutions to identify shapes. Finally, it uses two dense layers to classify the images into a vector of 64 different classes. The size of the images determines the complexity of the network. The larger the images, the more convolutions are required, and the larger the first layer. Larger networks take more time to train (the difference may be measured in days) and require more data (the difference may be measured in tens of thousands of images, depending on the number of classes and quality of images).

Therefore, for many applications, we use grayscale images and downsize them significantly. *Figure 3.6* shows the same image as previously but in grayscale, downsized to 192 x 108 pixels. The size of the image has been significantly reduced and so have the requirements for the first convolutional layers:

Figure 3.6 – Black-and-white transformed image (lower-quality lossy transformation illustrated on purpose)

However, the object in the image is still perfectly visible and can be used for further analysis. Therefore, here is the next best practice.

> **Best practice #15**
> Downsize the size of your images and use as few colors as possible to reduce the computational complexity of your system.

Before designing the system, we need to understand what kinds of images we have and how we can use them. Then, we can perform these kinds of transformations so that the system that we design can handle the tasks that it is designed for. However, it's important to note that downsizing images can also result in loss of information, which can affect the accuracy of the ML model. It's important to carefully balance the trade-offs between computational complexity and information loss when deciding how to preprocess images for an ML task.

Downsizing and converting images to grayscale is a common practice in ML. In fact, there exist several well-known and widely used benchmark datasets that use this technique. One of them is the MNIST dataset of handwritten numbers. The dataset is available for download as part of the most popular ML libraries such as TensorFlow and Keras. Just use the following code to get hold of the images:

```
# import the Keras library that contains the MNIST dataset
from keras.datasets import mnist
```

```
# load the dataset directly from the Keras website
# and use the standard train/test splits
(X_train, Y_train), (X_test, Y_test) = mnist.load_data()

# import a Matplot library to plot the images
from matplotlib import pyplot

# plot first few images
for i in range(9):
# define subplot to be 330 pixels wide
pyplot.subplot(330 + 1 + i)

# plot raw pixel data
   pyplot.imshow(X_train[i],
                  cmap=pyplot.get_cmap('gray'))

# show the figure
pyplot.show()
```

The code illustrates how to download the dataset, which is already split into test and train data, with annotations. It also shows how to visualize the dataset, which results in the images seen in *Figure 3.7*:

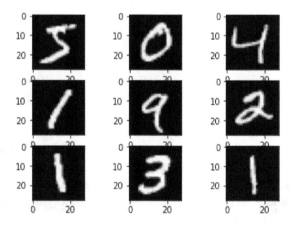

Figure 3.7 – Visualization of the first few images in the MNIST dataset; the
images are rasterized on purpose to illustrate their real size

The size of the images in the MNIST dataset is 28 x 28 pixels, which is perfectly sufficient to train and test new ML models. Although the dataset is well known and used in ML, it is relatively small and uniform – only grayscale numbers. Therefore, for more advanced tasks, we should look for more diverse datasets.

Images of handwritten numbers are naturally useful, but we often want to use more complex images, and therefore, the standard libraries contain images with more than just 10 classes (number of digits). One such dataset is the Fashion-MNIST dataset.

We can download it by using the following code:

```
from keras.datasets import fashion_mnist

(X_train, Y_train), (X_test, Y_test)=fashion_mnist.load_data()
```

The code, which we can use as part of the visualization shown in *Figure 3.7*, produces the set of images seen in *Figure 3.8*. The images are of the same size and the same number of classes, but with a larger complexity:

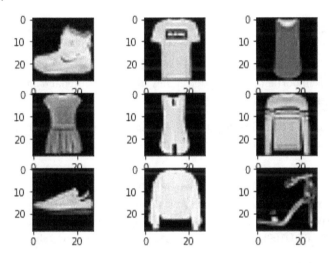

Figure 3.8 – Fashion-MNIST dataset; the images are rasterized on purpose to illustrate their real size

Finally, we can also use libraries that contain color images, such as the CIFAR-10 dataset. The dataset can be accessed with the following code:

```
from keras.datasets import cifar10
# load dataset
(X_train, Y_train), (X_test, Y_test)== cifar10.load_data()
```

The dataset contains images of 10 different classes of similar size (32 x 32 pixels), but with colors, as shown in *Figure 3.9*.

Figure 3.9 – CIFAR-10 dataset; the images are rasterized on purpose to illustrate their real size

This is not the end of such benchmark datasets. Some datasets contain more classes, larger images, or both. Therefore, it is important to take a look at these datasets beforehand and in connection with the task that our system has to perform.

In the majority of cases, grayscale images are perfectly fine for classification tasks. They provide the ability to quickly get an orientation in the data, and they are small enough that the classification is of good quality.

The usual size of the benchmark datasets is about 50,000–100,000 images. This illustrates that even for such a small number of classes and for such small images, the number is significant. Just imagine annotating those 100,000 images. For more complex images, the size of the datasets can be significantly larger. For example, the BDD100K dataset used in automotive software contains over 100,000 images.

Therefore, here is my next best practice.

> **Best practice #16**
>
> Use a reference dataset (such as MNIST or STL) for benchmarking whether the system works or not.

In order to understand whether the entire system works or not, such benchmark datasets are very useful. They provide us with a preconfigured train/test split, and there are plenty of algorithms that can be used to understand the quality of our algorithm.

We should also consider the next best practice.

> **Best practice #17**
> Whenever possible, use models that are already pre-trained for specific tasks (for example, neural models for image classification or semantic segmentation).

Just as we should strive to reuse images for benchmarking, we should also strive to reuse models that are pre-trained. This saves previous design resources and reduces the risk of spending too much time to find optimal architectures of NN models or to find the optimal set of parameters (even if we use `GradientSearch` algorithms).

Text

One of the types of analysis done on text is SA – classification of whether a piece of text (a sentence, for example) is positive or negative.

Figure 3.10 presents an example of data that can be used for SA. The data is publicly available and has been created from Amazon product reviews. Data for this kind of analysis is often structured in a table, where we have entities such as `ProductId` (I've truncated the `Id` columns for brevity) or `UserId`, as well as `Score` for reference and `Text` to classify.

This data structure provides us with the possibility to quickly summarize the text and visualize it. The visualization can be done in several ways – for example, by plotting a histogram of the scores. However, the most interesting visualizations are the ones that are provided by the statistics of words/tokens used in the text:

Id	ProductId	UserId	Score	Summary	Text
1	B001KFG0	UHU8GW	5	Good Quality Dog Food	I have bought several of the Vitality canned dog food products and have found them all to be of good quality. The product looks more like a stew than a processed meat and it smells better. My Labrador is finicky and she appreciates this product better than most.
2	B008GRG4	ZCVE5NK	1	Not as Advertised	Product arrived labeled as Jumbo Salted Peanuts...the peanuts were actually small sized unsalted. Not sure if this was an error or if the vendor intended to represent the product as "Jumbo".

Id	ProductId	UserId	Score	Summary	Text
3	B000OCH0	WJIXXAIN	4	"Delight" says it all	This is a confection that has been around a few centuries. It is a light, pillowy citrus gelatin with nuts - in this case Filberts. And it is cut into tiny squares and then liberally coated with powdered sugar. And it is a tiny mouthful of heaven. Not too chewy, and very flavorful. I highly recommend this yummy treat. If you are familiar with the story of C.S. Lewis' "The Lion, The Witch, and The Wardrobe" - this is the treat that seduces Edmund into selling out his Brother and Sisters to the Witch.
4	B000A0QIQ	C6FGVXV	2	Cough Medicine	If you are looking for the secret ingredient in Robitussin I believe I have found it. I got this in addition to the Root Beer Extract I ordered (which was good) and made some cherry soda. The flavor is very medicinal.
5	B0062ZZ7K	LF8GW1T	5	Great taffy	Great taffy at a great price. There was a wide assortment of yummy taffy. Delivery was very quick. If your a taffy lover, this is a deal.

Figure 3.10 – Example of data for product reviews, structured for SA; only the first five rows are shown

One way to visualize the data is to use the word cloud visualization technique. A simple script for visualizing this kind of data is shown next:

```
# Create stopwords list:
stopwords = set(STOPWORDS)
stopwords.update(["br", "href"])


# create text for the wordcloud
textt = " ".join(review for review in dfRaw.Text)


# generation of wordcloud
wordcloud = WordCloud(stopwords=stopwords,
                max_words=100,
                background_color="white").generate(textt)


# showing the image
```

```
# and saving it to the png file
plt.figure(figsize = [12,9])
plt.imshow(wordcloud, interpolation='bilinear')
plt.axis("off")
plt.show()
```

The result of running this script is shown in *Figure 3.11*. A word cloud shows trends in terms of frequency of the use of words – words used more frequently are larger than words used less frequently:

Figure 3.11 – Word cloud visualization of the Text column

Hence, my next best practice is as follows.

> **Best practice #18**
>
> Visualize your raw data to get an understanding of patterns in your data.

Visual representation of data is important to understand the underlying patterns. I cannot stress that enough. I use both Python's Matplotlib and Seaborn as well as visual analytics tools such as TIBCO Spotfire to plot charts and understand my data. Without such a visualization, and thus without such an understanding of the patterns, we are bound to make wrong conclusions and even design systems with flaws that are difficult to remove without a complete redesign.

Visualization of output from more advanced text processing

Visualization of text helps us to understand what the text contains, but it does not capture the meaning of it. In this book, we will work with advanced text processing algorithms – feature extractors. Therefore, we need to understand how to create visualizations of the output from these algorithms.

One way of working with feature extraction is to use word embeddings – a method to convert words or sentences into vectors of numbers. word2vec is one model that can do that, but there are more powerful ones too. OpenAI's GPT-3 model is one of the largest models that are openly available. Obtaining embeddings of paragraphs is quite straightforward. First, we connect to the OpenAI API and then query it for the embeddings. Here is the code that does the querying of the OpenAI API (in boldface):

```
# first we combine the title and the content of the review
df['combined'] = "Title: " + df.Summary.str.strip() + "; Content: " +
df.Text.str.strip()

# we define a function to get embeddings, to make the code more
straightforward
def get_embedding(text, engine="text-similarity-davinci-001"):
    text = text.replace("\n", " ")
    return openai.Embedding.create(input = [text], engine=engine)
['data'][0]['embedding']

# and then we get embeddings for the first 5 rows
df['babbage_similarity'] = df.head(5).combined.apply(lambda x: get_
embedding(x, engine='text-similarity-babbage-001'))
```

What we obtain by running this piece of code is 5 vectors (one for each row) of 2048 numbers, which we call embeddings. The entire vector is too large to be included in the page, but the first elements look something like this: [-0.005302980076521635, 0.018141526728868484, -0.018141526728868484, 0.004692177753895521, ...

These numbers do not say much to us humans, but they have a meaning for the language model. The meaning is in the distance between them – words/tokens/sentences that are like one another are placed closer than words/tokens/sentences that are not similar. In order to understand these similarities, we use transformations that reduce the dimensions – one of them is **t-distribution Stochastic Neighbor Embedding** (**t-SNE**). *Figure 3.12* presents this kind of visualization of the five embeddings that we obtained:

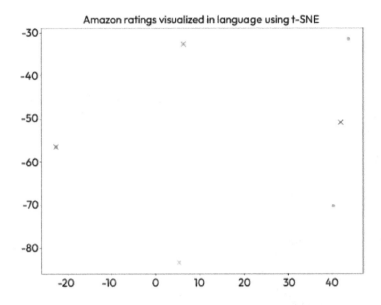

Figure 3.12 – t-SNE visualization of the embedding vectors for the five reviews

Each dot represents one review, and each cross represents the center of the cluster. The clusters are designated by the `Score` column from the original dataset. The screenshot shows that the text in each of the reviews is different (dots do not overlap) and that the clusters are separate – the crosses are positioned in a different part of the screenshot.

Therefore, my next best practice is about that.

Best practice #19

Visualize your data when it has been turned into features to monitor whether the same patterns are still observable.

Just as with the previous best practice, we need to visualize the data to check whether patterns that we observed in the raw data are still observable. This step is important as we need to know that the ML model can indeed learn these patterns. As these models are statistical in nature, they always capture patterns, but when a pattern is not there, they capture something that is not useful – even though it resembles a pattern.

Structured text – source code of programs

The source code of programs is a special case of text data. It has the same type of modality – text – but it contains additional information in the form of the grammatical/syntactical structure of the program. Since every programming language is based on grammar, there are specific rules for how a program should be structured. For instance, in C, there should be a specific function called `main`, which is the entry point for the program.

These specific rules make the text structured in a specific way. They may make it more difficult to understand the text for a human being, but this structure can certainly be helpful. One of the models that uses this structure is `code2vec` (`https://code2vec.org/`). The `code2vec` model is similar to word2vec, but it takes as input the **Abstract Syntax Tree** (**AST**) of the program that it analyzes – for example, the following program:

```
void main()
{
    Console.println("Hello World");
}
```

This can be represented by the AST in *Figure 3.13*:

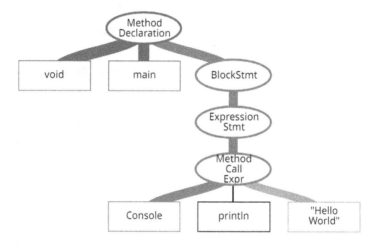

Figure 3.13 – AST for a simple "Hello World" program

The example program is visualized as a set of instructions with their context and the role that they play in the program. For example, the words `void` and `main` are two parts of the method declaration, together with the block statement (`BlockStmt`), which is the body of the method.

`code2vec` is an example of a model that uses programming language information (in this case, grammar) as input to the model. Tasks that the model can do include finding similarities between words (such as word2vec models), finding combinations, and identifying analogies. For example, the model can identify all combinations of the words `int` and `main` and provide the following answers (with probabilities): `realMain` (71%), `isInt` (71%), and `setIntField` (69%). These tasks, by extension, can be used for program repair, where the model can identify mistakes and repair them.

However, using an AST or similar information has disadvantages. The main disadvantage is that the analyzed program must compile. This means that we cannot use these kinds of models in the context where we want to analyze programs that are incomplete – for example, in the context of **continuous integration** (**CI**) or modern code reviews. When we analyze only a small part of the code, the model cannot parse it, obtain its AST, and use it. Therefore, here's my next best practice.

> **Best practice #20**
>
> Only use the necessary information as input to ML models. Too much information may require additional processing and make the training hard to converge (finish).

When designing the processing pipeline, make sure that the information provided to the model is necessary, as every piece of information poses new requirements for the entire software system. As in the example of an AST, when it is necessary, it is powerful information, but if not available, it can be a huge hindrance to getting the data analysis pipeline to work.

Every data has its purpose – annotations and tasks

Data in raw format is important, but only the first step in the development and operations of ML software. The most important part, and the costliest one, is the annotation of the data. To train an ML model and then use it to make inferences, we need to define a task. Defining a task is both conceptual and operational. The conceptual definition is to define what we want the software to do, but the operational definition is how we want to achieve that goal. The operational definition boils down to a definition of what we see in the data and what we want the ML model to identify/replicate.

Annotations are the mechanisms by which we direct the ML algorithms. Every piece of data that we use requires some sort of label to denote what it is. In the raw format of the data, this annotation can be a label of what the data point contains. For example, such a label can be that the image contains the number 1 (from the MNIST dataset) or a car (from the CIFAR-10 dataset). However, these simple annotations are important only in dedicated tasks. For more advanced tasks, the annotations need to be richer.

One of the types of annotations relates to when we designate part of the data as interesting. In the case of images, this is done by drawing bounding boxes around objects of interest. *Figure 3.14* presents such an image:

Figure 3.14 – Image with bounding boxes

The image contains boxes around elements that we want the model to recognize. In this case, we want to recognize objects that are vehicles (green boxes), other road users (orange boxes), and important background objects (gray boxes). These kinds of annotations are used for the ML model to learn shapes and to identify such shapes in new objects. In the case of this example, the bounding boxes identify elements that are important for active safety systems in cars, but this is not the only application.

Other applications of such bounding boxes include medical image analysis, where the task is to identify tissue that needs further analysis. These could also be systems for face recognition and object detection.

Although this task, and the bounding boxes, could be seen as a special case of annotating raw data, it is a bit different. Every box could be seen as a separate image that has a label, but the challenge is that every box is of a different size. Therefore, using such differently shaped images would require preprocessing (for example, rescaling). It would also only work for the training because, in the inference, we would need to identify objects before they are classified – and that's exactly the task we need the NN to do for us.

My best practice for using this kind of data is set out next.

> **Best practice #21**
>
> Use bounding boxes in the data when the task requires the detection and tracking of objects.

Since bounding boxes allow us to identify objects, the natural use of this data is in tracking systems. An example application is a system that monitors parking spots using a camera. It detects parking spots and tracks whether there is a vehicle parked in the spot or not.

An extension of the object detection task is a perception task, where our ML software needs to make decisions based on the context of the data – or a situation.

For image data, this context can be described by a semantic map. *Figure 3.15* shows such a map:

Figure 3.15 – Image with a semantic map; building is one of the labels for the semantic map

The screenshot shows an overlay of different colors that cover objects of specific types. The orange overlay shows vehicles, while the purple one shows vulnerable road users, which is the pedestrian in this image. Finally, the pink color indicates buildings, and the red color covers the background/sky.

The semantic map provides more flexibility than bounding boxes (as some objects are more interesting than others) and allows the ML system to get the context of the image. By identifying what kinds of elements are present in the image, the ML model can provide information about where the image was taken to the decision algorithm of the software system that we design.

Therefore, here's my next best practice.

> **Best practice #22**
> Use semantic maps when you need to get the context of the image or you need details of a specific area.

Semantic maps require heavy computations to be used effectively; therefore, we should use them scarcely. We should use these maps when we have tasks related to the context, such as perception algorithms or image alteration – for example, changing the color of the sky in the image. Regarding the accuracy of the information, it is generally true that semantic maps require heavy computations and are therefore used selectively. An example of a tool that does such a semantic mapping is Segments.ai.

Semantic maps are useful when the task requires understanding the context of an image or details of a specific area. For example, in autonomous driving, a semantic map can be used to identify objects on the road and their relationship to each other, allowing the vehicle to make informed decisions about its movement. However, specific use cases for semantic maps may vary depending on the application.

Annotating text for intent recognition

SA, which we mentioned before, is only one type of annotation of textual data. It is useful for assessing whether the text is positive or negative. However, instead of annotating text with a sentiment, we can annotate the text with – for example – the intent and train an ML model to recognize intent from other text passages. The table in *Figure 3.16* provides such an annotation, based on the same review data as before:

Id	Score	Summary	Text	Intent
1	5	Good Quality Dog Food	I have bought several of the Vitality canned dog food products and have found them all to be of good quality. The product looks more like a stew than a processed meat and it smells better. My Labrador is finicky and she appreciates this product better than most.	Advertise
2	1	Not as Advertised	Product arrived labeled as Jumbo Salted Peanuts...the peanuts were actually small sized unsalted. Not sure if this was an error or if the vendor intended to represent the product as "Jumbo".	Criticize
3	4	"Delight" says it all	This is a confection that has been around a few centuries. It is a light, pillowy citrus gelatin with nuts - in this case Filberts. And it is cut into tiny squares and then liberally coated with powdered sugar. And it is a tiny mouthful of heaven. Not too chewy, and very flavorful. I highly recommend this yummy treat. If you are familiar with the story of C.S. Lewis' "The Lion, The Witch, and The Wardrobe" - this is the treat that seduces Edmund into selling out his Brother and Sisters to the Witch.	Describe
4	2	Cough Medicine	If you are looking for the secret ingredient in Robitussin I believe I have found it. I got this in addition to the Root Beer Extract I ordered (which was good) and made some cherry soda. The flavor is very medicinal.	Criticize
5	5	Great taffy	Great taffy at a great price. There was a wide assortment of yummy taffy. Delivery was very quick. If your a taffy lover, this is a deal.	Advertise

Figure 3.16 – Textual data annotation for intent recognition

The last column of the table shows the annotation of the text – the intent. Now, we can use the intent as a label to train an ML model to recognize intent in the new text. Usually, this task requires a two-step approach, as shown in *Figure 3.17*:

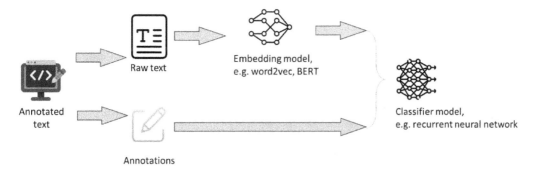

Figure 3.17 – A two-step approach for training models based on text

The annotated text is organized into two parts. The first part is the text itself (for example, the `Text` column in our example table), and the second part is the annotation (e.g., the `Intent` column in our example). The text is processed using a model such as the word2vec model or a transformer, which encodes the text as a vector or a matrix. The annotations are encoded into a vector using techniques such as one-hot encoding so that they can be used as decision classes for the classification algorithm. The classification algorithm takes both the encoded annotations and the vectorized text. Then, it is trained to find the best fit of the vectorized text (X) to the annotation (Y).

Here is my best practice on how to perform that.

> **Best practice #23**
> Use a pre-trained embedding model such as GPT-3 or an existing BERT model to vectorize your text.

From my experience, working with text is often easier if we use a predefined language model to vectorize the text. The *Hugging Face* website (`www.huggingface.com`) is an excellent source of these models. Since LLMs require significant resources to train, the existing models are often good enough for most of the tasks. Since we develop the classifier model as the next step in the pipeline, we can focus our efforts on making that model better and align it with our task.

Another type of annotation of text data is the context in terms of the **part of speech** (**POS**). It can be seen as the semantic map used in the image data. Each word is annotated, whether it is a noun, verb, or adjective, regardless of which part of the sentence it belongs to. An example of such an annotation can be presented visually, using the Allen Institute's AllenNLP **Semantic Role Labeling** (**SRL**) tool (`https://demo.allennlp.org/semantic-role-labeling`). *Figure 3.18* presents a screenshot of such labeling for a simple sentence, while *Figure 3.19* presents the labeling for a more complex one:

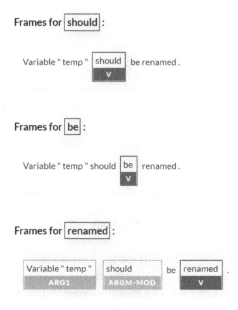

Figure 3.18 – SRL using the AllenNLP toolset

In this sentence, the role of each word is emphasized, and we see that there are three verbs with different associations – the last one is the main one as it links the other parts of the sentence:

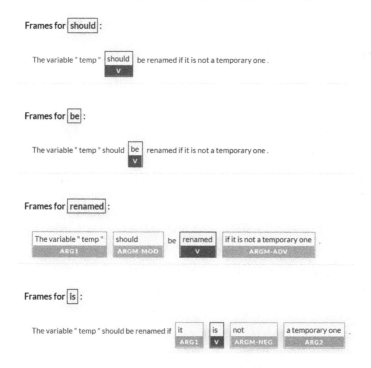

Figure 3.19 – SRL for a more complex sentence

The complex sentence has a larger semantic role frame, as it contains two distinct parts of the sentence. We use this role labeling in order to extract the meaning of passages of text. It is particularly useful when designing software systems based on so-called *grounded models*, which are models that check the information toward a ground truth. Such models parse the text data, find the right anchor (for example, what the question is about), and find the relevant answer in their database. These are opposed to *ungrounded models*, which create answers based on which word fits best to finish the sentence – for example, ChatGPT.

Therefore, my best practice is shown next.

> **Best practice #24**
> Use role labels when designing software that needs to provide grounded decisions.

Grounded decisions are often more difficult to provide, as the model needs to understand the context of the sentence, capture its meaning, and provide relevant answers. However, this is not always needed or even desired. Ungrounded models are often good enough for suggestions that can be later fixed by specialists. An example of such a software tool is ChatGPT, which provides answers that are sometimes incorrect and require manual intervention. However, they are a very good start.

Where different types of data can be used together – an outlook on multi-modal data models

This chapter introduced three types of data – images, text, and structured text. These three types of data are examples of data that is in a numerical form, such as matrices of numbers, or in forms of time series. Regardless of the form, however, working with data and ML systems is very similar. We need to extract the data from a source system, then transform it into a format that we can annotate, and then use this as input to an ML model.

When we consider different types of data, we could start to think about whether we could use two types of data in the same system. There are a few ways of doing that. The first one is when we use different ML systems in different pipelines, but we connect the pipelines. GitHub Copilot is such a system. It uses a pipeline for processing a natural language to find similar programs and to transform them so that they fit the context of the program being developed now.

Another example is a system that generates textual descriptions of images. It takes an image as an input, identifies objects in it, and then generates text based on these objects. The generation of text is done by a completely different ML model than the classification of images.

However, there are new models that use two different modalities – images and texts – in the same NN – for example, the Gato model. By using the input from two sources and using a very narrow (in terms of the number of neurons) network in the middle, the model is trained to generalize concepts described by two different modalities. In this way, the model is trained to understand that an image of a cat and the word "cat" should be placed in the same embedding space very close to one another, if not exactly in the same places. Although still experimental, these kinds of networks are intended to mimic the human understanding of concepts.

In the next chapter, we will dive a bit deeper into the understanding of data by making a deeper dive into the process of feature engineering.

References

- Tao, J. et al., An object detection system based on YOLO in traffic scene. In 2017 6th International Conference on Computer Science and Network Technology (ICCSNT). 2017. IEEE.

- Artan, C.T. and T. Kaya, Car Damage Analysis for Insurance Market Using Convolutional Neural Networks. In International Conference on Intelligent and Fuzzy Systems. 2019. Springer.

- Nakaura, T. et al., A primer for understanding radiology articles about machine learning and deep learning. Diagnostic and Interventional Imaging, 2020. 101(12): p. 765-770.

- Bradski, G., The OpenCV Library. Dr. Dobb's Journal: Software Tools for the Professional Programmer, 2000. 25(11): p. 120-123.

- *Memon, J. et al., Handwritten optical character recognition (OCR): A comprehensive systematic literature review (SLR). IEEE Access, 2020. 8: p. 142642-142668.*

- *Mosin, V. et al., Comparing autoencoder-based approaches for anomaly detection in highway driving scenario images. SN Applied Sciences, 2022. 4(12): p. 1-25.*

- *Zeineldin, R.A. et al., DeepSeg: deep neural network framework for automatic brain tumor segmentation using magnetic resonance FLAIR images. International journal of computer assisted radiology and surgery, 2020. 15(6): p. 909-920.*

- *Reid, R. et al., Cooperative multi-robot navigation, exploration, mapping and object detection with ROS. In 2013 IEEE Intelligent Vehicles Symposium (IV). 2013. IEEE.*

- *Mikolov, T. et al., Recurrent neural network based language model. In Interspeech. 2010. Makuhari.*

- *Vaswani, A. et al., Attention is all you need. Advances in neural information processing systems, 2017. 30.*

- *Ma, L. and Y. Zhang, Using Word2Vec to process big text data. In 2015 IEEE International Conference on Big Data (Big Data). 2015. IEEE.*

- *Ouyang, X. et al., Sentiment analysis using convolutional neural network. In 2015 IEEE International Conference on Computer and Information Technology; ubiquitous computing and communications; dependable, autonomic and secure computing; pervasive intelligence and computing. 2015. IEEE.*

- *Roziere, B. et al., Unsupervised translation of programming languages. Advances in Neural Information Processing Systems, 2020. 33: p. 20601-20611.*

- *Yasunaga, M. and P. Liang, Break-it-fix-it: Unsupervised learning for program repair. In International Conference on Machine Learning. 2021. PMLR.*

- *Halali, S. et al., Improving defect localization by classifying the affected asset using machine learning. In International Conference on Software Quality. 2019. Springer.*

- *Ochodek, M. et al., Recognizing lines of code violating company-specific coding guidelines using machine learning. In Accelerating Digital Transformation. 2019, Springer. p. 211-251.*

- *Nguyen, N. and S. Nadi, An empirical evaluation of GitHub copilot's code suggestions. In Proceedings of the 19th International Conference on Mining Software Repositories. 2022.*

- *Zhang, C.W. et al., Pedestrian detection based on improved LeNet-5 convolutional neural network. Journal of Algorithms & Computational Technology, 2019. 13: p. 1748302619873601.*

- *LeCun, Y. et al., Gradient-based learning applied to document recognition. Proceedings of the IEEE, 1998. 86(11): p. 2278-2324.*

- *Xiao, H., K. Rasul, and R. Vollgraf, Fashion-MNIST: a novel image dataset for benchmarking machine learning algorithms. arXiv preprint arXiv:1708.07747, 2017.*

- *Recht, B. et al., Do CIFAR-10 classifiers generalize to CIFAR-10? arXiv preprint arXiv:1806.00451, 2018.*

- *Robert, T., N. Thome, and M. Cord, HybridNet: Classification and reconstruction cooperation for semi-supervised learning. In Proceedings of the European Conference on Computer Vision (ECCV). 2018.*

- *Yu, F. et al., Bdd100k: A diverse driving video database with scalable annotation tooling. arXiv preprint arXiv:1805.04687, 2018. 2(5): p. 6.*

- *McAuley, J.J. and J. Leskovec, From amateurs to connoisseurs: modeling the evolution of user expertise through online reviews. In Proceedings of the 22nd International Conference on World Wide Web. 2013.*

- *Van der Maaten, L. and G. Hinton, Visualizing data using t-SNE. Journal of Machine Learning Research, 2008. 9(11).*

- *Sengupta, S. et al., Automatic dense visual semantic mapping from street-level imagery. In 2012 IEEE/RSJ International Conference on Intelligent Robots and Systems. 2012. IEEE.*

- *Palmer, M., D. Gildea, and N. Xue, Semantic role labeling. Synthesis Lectures on Human Language Technologies, 2010. 3(1): p. 1-103.*

- *Reed, S. et al., A generalist agent. arXiv preprint arXiv:2205.06175, 2022.*

4

Data Acquisition, Data Quality, and Noise

Data for machine learning systems can come directly from humans and software systems – usually called *source systems*. Where the data comes from has implications regarding what it looks like, what kind of quality it has, and how to process it.

The data that originates from humans is usually noisier than data that originates from software systems. We, as humans, are known for small inconsistencies and we can also understand things inconsistently. For example, the same defect reported by two different people could have a very different description; the same is true for requirements, designs, and source code.

The data that originates from software systems is often more consistent and contains less noise or the noise in the data is more regular than the noise in the human-generated data. This data is generated by source systems. Therefore, controlling and monitoring the quality of the data that's generated automatically is different – for example, software systems do not "lie" in the data, so it makes no sense to check the believability of the data that's generated automatically.

In this chapter, we'll cover the following topics:

- The different sources of data and what we can do with them

- How to assess the quality of data that's used for machine learning

- How to identify, measure, and reduce noise in data

Sources of data and what we can do with them

Machine learning software has become increasingly important in all fields today. Anything from telecommunication networks, self-driving vehicles, computer games, smart navigation systems, and facial recognition to websites, news production, cinematography, and experimental music creation can be done using machine learning. Some applications are very successful at, for example, using machine learning in search strings (BERT models). Some applications are not so successful, such as using machine learning in hiring processes. Often, this depends on the programmers, data scientists, or models that are used in these applications. However, in most cases, the success of a machine learning application is often in the data that is used to train it and use it. It depends on the quality of that data and the features that are extracted from it. For example, Amazon's machine learning recommender was taken out of operation because it was biased against women. Since it was trained on historical recruitment data, which predominantly included male candidates, the system tended to recommend male candidates in future hiring.

Data that's used in machine learning systems can originate from all kinds of sources. However, we can classify these sources into two main types – manual/human and automated software/hardware. These two types have different characteristics that determine how to organize these sources and how to extract features from the data from these sources. *Figure 4.1* illustrates these types of data and provides examples of data of each type.

Manually generated data is the data that comes from human input or originates in humans. This kind of data is often much richer in information than the data generated by software, but it also has much larger variability. This variability can come from the natural variability of us, humans. The same question in a form, for example, can be understood and answered differently by two different people. This kind of data is often more susceptible to random errors than systematic ones.

Automatically generated data originates from hardware or software systems, usually measured by sensors or measurement scripts that collect data from other systems, products, processes, or organizations. This kind of data is often more consistent and repeatable but also more susceptible to systematic errors:

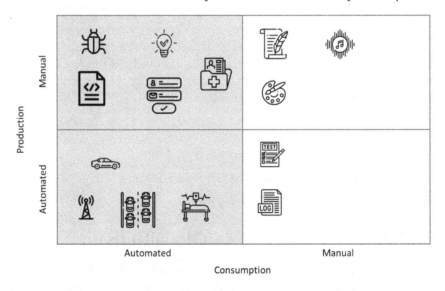

Figure 4.1 – Sources of data and their classification. The green part of this figure is the scope of this book

An example of data that originates from humans, and is often used in machine learning, is data about software defects. A human tester often inspects a problem and reports it using a form. This form contains fields about the testing phase, components affected, impact on the customer, and more, but one part of the form is often a description of the problem in natural language, which is an interpretation of what happens when it's authored by a human tester.

Another type of data that's generated by humans is source code. We, as programmers, write source code for the software in a programming language with a given syntax, and we use programming guidelines to keep a consistent style so that the product of our work – the source code – can be automatically interpreted or compiled by software. There are some structures in the code that we write, but it is far from being consistent. Even the same algorithm, when implemented by two different programmers, differs in the naming of variables, their types, or even how the problem is solved. A good example of this is the Rosetta code website, which provides the same solutions in different programming languages and sometimes even in the same programming language.

Requirement specifications or data input using forms have the same properties and characteristics.

However, there is one origin of data that is particularly interesting – medical data. This is the data from patients' records and charts, input by the medical specialists as part of medical procedures. This data can be electronic, but it reflects the specialists' understanding of the symptoms and their interpretation of the medical tests and diagnoses.

On the other hand, we have data that's generated by software or hardware in one way or another. The data that's generated automatically is more consistent, although not free from problems, and more repetitive. An example of such data is the data generated in telecommunication networks to transmit information from one telecommunication node to another. The radio signals are very stable compared to other types of data, and can be disturbed by external factors such as weather (precipitation) or obstructions such as construction cranes. The data is repeatable, with all variability originating from the external factors.

Another example is the data from vehicles, which register information around them and store it for further processing. This data can contain signaling between the vehicle's components as well as communication with other vehicles or the infrastructure.

Medical data, such as **electroencephalograph** (**EEG** – that is, brain waves) or **electrocardiogram** (**ECG** – that is, heart rate), is collected from source systems, which we can see as measurement instruments. So, technically, it has been generated by computer systems, but it comes from human patients. This origin in patients means that the data has the same natural variability as any other data collected from humans. As every patient is a bit different, and the measurement systems can be attached to each patient a bit differently, the data generated by each patient differs slightly from other patients. For example, the ECG heartbeat data contains basic, consistent information – the number of beats per minute (among other parameters). However, the raw data differs in the amplitude of the ECG signal (depending on the placement of the measurement electrode) or the difference between spikes in the curve (R and T-spikes).

Therefore, my first best practice in this chapter has to do with the origin of the data that we use for software systems.

> **Best practice #25**
> Identify the origin of the data used in your software and create your data processing pipeline accordingly.

Since all types of data require different ways of working in terms of cleaning, formatting, and feature extraction, we should make sure that we know how the data is produced and what kind of problems we can expect (and handle). Therefore, first, we need to identify what kind of data we need, where it comes from, and what kind of problems it can carry with it.

Extracting data from software engineering tools – Gerrit and Jira

To illustrate how to work with data extraction, let's extract data from a popular software engineering tool for code reviews – Gerrit. This tool is used for reviewing and discussing fragments of code developed by individual programmers, just before they are integrated into the main code base of the product.

The following program code shows how to access the database of Gerrit – that is, through the JSON API – and how to extract the list of all changes for a specific project. This program uses the Python `pygerrit2` package (`https://pypi.org/project/pygerrit2/`). This module helps us use the JSON API as it provides Python functions instead of just JSON strings:

```
# importing libraries
from pygerrit2 import GerritRestAPI
# A bit of config - repo
gerrit_url = "https://gerrit.onap.org/r"
# since we use a public OSS repository
auth = None


# this line gets sets the parameters for the HTML API
rest = GerritRestAPI(url=gerrit_url, auth = auth)


# the main query where we ask the endpoint to provide us the list and
details of all changes
# each change is essentially a review that has been submitted to the
repository
changes = rest.get("/changes/?q=status:merged&o=ALL_FILES&o=ALL_
REVISIONS&o=DETAILED_LABELS&start=0",
headers={'Content-Type': 'application/json'})
```

The key line in this code fragment is `rest.get("/changes/?q=status:merged&o=ALL_FILES&o=ALL_REVISIONS&o=DETAILED_LABELS&start=0", headers={'Content-Type': 'application/json'})`. This line specifies the endpoint for the changes to be retrieved, along with the parameters. It says that we want to access all files and all revisions, as well as all details of the changes. In these details, we can find the information about all revisions (particular patches/commits) and we can then parse each of these revisions. It is important to know that the JSON API returns a maximum of 500 changes in each query, and therefore, the last parameter – `start=0` – can be used to access changes from 500 upward. The output of this program is a very long list of changes in JSON, so I will not present it in detail in this book. Instead, I encourage you to execute this script and go through this file at your own pace. The script can be found in this book's GitHub repository at `https://github.com/miroslawstaron/machine_learning_best_practices`, under `chapter 4`. The name of the script is `gerrit_exporter.ipynb`.

Now, extracting just the list of changes is not very useful for analyses as it only provides the information that has been collected automatically – for example, which revisions exist, and who created these revisions. It does not contain information about which files and lines are commented on, or what the comments are – in other words, the particularly useful information. Therefore, we need to interact with Gerrit a bit more, as presented in *Figure 4.2*.

The program flow presented in *Figure 4.2* illustrates the complexity of the relationships in a code comment database such as Gerrit. Therefore, the program to access the database and export this data is a bit too long for this book. It can be accessed in the same repository as the previous one, under the name `gerrit_exporter_loop.ipynb`:

Access Gerrit JSON API

Get list of all changes

Get list of all revisions for each change

Get list of all comments for each file

Get list of all lines that are commented

Figure 4.2 – Interactions with Gerrit to extract commented files,
commented lines, and the content of the comments

This kind of data can be used to train machine learning models to review the code or even identify which lines of code need to be reviewed.

When working with Gerrit, I have found the following best practice to be useful.

> **Best practice # 26**
>
> Extract as much data as you need and store it locally to reduce the disturbances for the software engineers who use the tool for their work.

Although it is possible to extract the changes one by one, it is better to extract the whole set of changes once and keep a local copy of it. In this way, we reduce the strain on the servers that others use for their daily work. We must remember that data extraction is a secondary task for these source systems, while their primary task is to support software engineers in their work.

Another good source of data for software systems that support software engineering tasks is JIRA, an issue and task management system. JIRA is used to document epics, user stories, software defects, and tasks and has become one of the most popular tools for this kind of activity.

Therefore, we can extract a lot of useful information about the processes from JIRA as a source system. Then, we can use this information to develop machine learning models to assess and improve tasks and requirements (in the form of user stories) and design tools that can help us identify overlapping user stories or group them into more coherent epics. Such software can be used to improve the quality of these tasks or provide better estimations.

The following code fragment illustrates how to make a connection to a JIRA instance and then how to extract all issues for a particular project:

```
# import the atlassian module to be able to connect to JIRA
from atlassian import Jira
jira_instance = Jira(
    #Url of the JIRA server
    url="https://miroslawstaron.atlassian.net/",
    #  user name
    username='email@domain.com',
    # token
    password='your_token',
    cloud=True
)
# get all the tasks/ issues for the project
jql_request = 'project = MLBPB'
issues = jira_instance.jql(jql_request)
```

In this fragment, for illustration purposes, I'm using my own JIRA database and my own project (MLBPB). This code requires the `atlassian-python-api` module to be imported. This module provides an API to connect to and interact with a JIRA database using Python, in a similar way as the API of Gerrit. Therefore, the same best practice as for Gerrit applies to JIRA.

Extracting data from product databases – GitHub and Git

JIRA and Gerrit are, to some extent, additional tools to the main product development tools. However, every software development organization uses a source code repository to store the main asset – the source code of the company's software product. Today, the tools that are used the most are Git version control and its close relative, GitHub. Source code repositories can be a very useful source of data for machine learning systems – we can extract the source code of the product and analyze it.

GitHub is a great source of data for machine learning if we use it responsibly. Please remember that the source code provided as open source, by the community, is not for profiting off. We need to follow the licenses and we need to acknowledge the contributions that were made by the authors, contributors, and maintainers of the open source community. Regardless of the license, we are always able to analyze our own code or the code that belongs to our company.

Once we can access the source code of our product or the product that we want to analyze, the following code fragment helps us connect to the GitHub server and access a repository:

```
# First create a Github instance:
# using an access token
g = Github(token, per_page=100)

# get the repo for this book
repo = g.get_repo("miroslawstaron/machine_learning_best_practices")

# get all commits
commits = repo.get_commits()
```

To support secure access to the code, GitHub uses access tokens rather than passwords when connecting to it. We can also use SSL and CLI interfaces, but for the sake of simplicity, we'll use the HTTPS protocol with a token. The `g = Github(token, per_page=100)` line uses the token to instantiate the PyGitHub library's main class. The token is individual and needs to be generated either per repository or per user.

The connection to the repository is done by the next line, `repo = g.get_repo("miroslawstaron/machine_learning_best_practices")`, which, in this example, connects to the repository associated with this book. Finally, the last line in the code fragment obtains the number of commits in the repository. Once obtained, we can print it and start analyzing it, as shown in the following code fragment:

```
# print the number of commits
print(f'Number of commits in this repo: {commits.totalCount}')
```

```
# print the last commit
print(f'The last commit message: {commits[0].commit.message}')
```

The last line of this code fragment prints out the commit message of the latest commit. It is worth noting that the latest commit is always first in the list of commits. Once we know the commit, we can also access the list of files that are included in that commit. This is illustrated by the following code fragment:

```
# print the names of all files in the commit
# 0 means that we are looking at the latest commit
print(commits[0].file)
```

Printing the list of the files in the commits is good, but it's not very useful. Something more useful is to access these files and analyze them. The following code fragment shows how to access two files from the latest two commits. First, we access the files, after which we download their content and store it in two different variables – linesOne and linesTwo:

```
# get one of the files from the commit
fileOne = commits[0].files[0]

# get the file from the second commit
fileTwo = commits[1].files[0]

# to get the content of the file, we need to get the sha of the commit
# otherwise we only get the content from the last commit
fl = repo.get_contents(fileOne.filename, ref=commits[0].sha)
fr = repo.get_contents(fileTwo.filename, ref=commits[1].sha)

# read the file content, but decoded into strings
# otherwise we would get the content in bytes
linesOne = fl.decoded_content
linesTwo = fr.decoded_content
```

Finally, we can analyze the two files for one of the most important tasks – to get the diff between the two files. We can use the difflib Python library for this task, as shown here:

```
# calculate the diff using difflib
# for which we use a library difflib
import difflib

# print diff lines by iterating the list of lines
# returned by the difflib library
for line in difflib.unified_diff(str(linesOne),
                                 str(linesTwo),
```

```
                                    fromfile=fileOne.filename,
                                    tofile=fileTwo.filename):
    print(line)
```

The preceding code fragment allows us to identify differences between files and print them in a way similar to how GitHub presents the differences.

My next best practice is related to the use of public repositories.

> **Best practice # 27**
>
> When accessing data from public repositories, please check the licenses and ensure you acknowledge the contribution of the community that created the analyzed code.

As I mentioned previously, open source programs are here for everyone to use, including to analyze and learn from them. However, the community behind this source code has spent countless hours creating and perfecting it. Therefore, we should use these repositories responsibly. If we use the repositories to create our own products, including machine learning software products, we need to acknowledge the community's contribution and, if we use the software under a copyleft license, give back our work to the community.

Data quality

When designing and developing machine learning systems, we consider the data quality on a relatively low level. We look for missing values, outliers, or similar. They are important because they can cause problems when training machine learning models. Nevertheless, they are nearly enough from a software engineering perspective.

When engineering reliable software systems, we need to know more about the data we use than whether it contains (or not) missing values. We need to know whether we can trust the data (whether it is believable), whether the data is representative, or whether it is up to date. So, we need a quality model for our data.

There are several quality models for data in software engineering, and the one I often use, and recommend, is the AIMQ model – a methodology for assessing information quality.

The quality dimensions of the AIMQ model are as follows (cited from Lee, Y.W., et al., *AIMQ: a methodology for information quality assessment*. Information & management, 2002. 40(2): p. 133-146):

- **Accessibility**: The information is easily retrievable and accessible to our system

- **Appropriate amount**: The information is of sufficient volume for our needs and for our applications

- **Believability**: The information is trustworthy and can be believed

- **Completeness**: The information includes all the necessary values for our system

- **Concise representation**: The information is formatted compactly and appropriately for our application

- **Consistent representation**: The information is consistently presented in the same format, including its representation over time

- **Ease of operation**: The information is easy to manipulate to meet our needs

- **Free of errors**: The information is correct, accurate, and reliable for the application that we're creating

- **Interpretability**: It is easy to interpret what the information means

- **Objectivity**: The information was objectively collected and is based on facts

- **Relevance**: The information is useful, relevant, appropriate, and applicable to our system

- **Reputation**: The information has a good reputation for quality and comes from good sources

- **Security**: The information is protected from unauthorized access and sufficiently restricted

- **Timeliness**: The information is sufficiently current for our work

- **Understandability**: The information is easy to understand and comprehend

Some of these dimensions are certainly universal for all kinds of applications. For example, the *free from error* dimension is relevant for all systems and all machine learning models. At the same time, *relevance* must be evaluated in the context of the application and software that we design. The dimensions that are closer to the raw data can be assessed automatically more easily than the dimensions related to applications. For application-related dimensions, we often need to make an expert assessment or conduct manual analyses or investigations.

Take believability, for example. To assess whether our source data is trustworthy and can be used for this application, we need to understand where the data comes from, who/what created that data, and under which premises. This cannot be automated as it requires human, expert, judgment.

Therefore, we can organize these dimensions at different abstraction levels – raw data or source systems, data for training and inference, machine learning models and algorithms, and the entire software product. It can be useful to show which of these dimensions are closer to the raw data, which ones are closer to the algorithms, and which ones are closer to the products. *Figure 4.3* shows this organization:

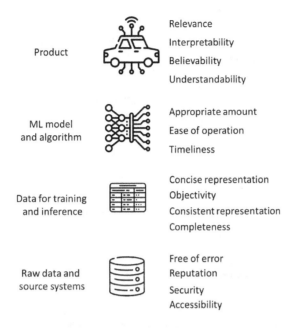

Product
- Relevance
- Interpretability
- Believability
- Understandability

ML model and algorithm
- Appropriate amount
- Ease of operation
- Timeliness

Data for training and inference
- Concise representation
- Objectivity
- Consistent representation
- Completeness

Raw data and source systems
- Free of error
- Reputation
- Security
- Accessibility

Figure 4.3 – Data quality attributes organized according to their logical relevance

Figure 4.3 indicates that there is a level of abstraction in how we check information quality, which is entirely correct. The lowest abstraction levels, or the first checks, are intended to quantify the basic quality dimensions. These checks do not have to be very complex either. The entire quality measurement and monitoring system can be quite simplistic, yet very powerful. *Figure 4.4* presents a conceptual design of such a system. This system consists of three parts – the machine learning pipeline, the log files, and the information quality measurement and monitoring:

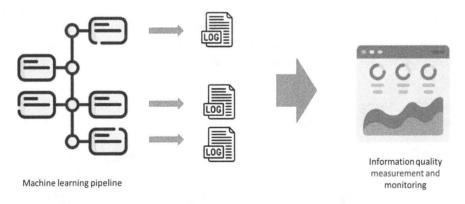

Machine learning pipeline

Information quality measurement and monitoring

Figure 4.4 – Information quality measurement system

First, the machine learning pipeline contains probes, or measurement instruments, that collect information about problems related to information quality. For instance, these instruments can collect information if there are any problems with accessing the data, which can indicate problems with the *accessibility* quality dimension.

The following code fragment shows how this can be realized in practice. It configures a rudimentary log file that collects the error information from the machine learning pipeline:

```
import logging


# create a logging file
# including the format of the log messages
logging.basicConfig(filename='./information_quality_gerrit.log',
                    filemode='w',
                    format='%(asctime)s;%(name)s;%(levelname)
s;%(message)s',
                    level=logging.DEBUG)


# specifying the name of the logger,
# which will tell us that the message comes from this program
# and not from any other modules or components imported
logger = logging.getLogger('Gerrit data export pipeline')


# the first log message to indicate the start of the execution
# it is important to add this, since the same log-file can be re-used
# the re-use can be done by other components to provide one single
point of logging
logger.info('Configuration started')
```

This code fragment creates log files, provides them with a unique name that the entire machine learning pipeline uses, and specifies the format of the error messages. Then, in the machine learning pipeline itself, the log files are propagated with messages. The following code fragment shows an example of how the data export tool presented previously is instrumented to propagate this information:

```
# A bit of config - repo
gerrit_url = "https://gerrit.onap.org/r"
fileName = "./gerrit_reviews.csv"


# since we use a public oss repository, we don't need to authenticate
auth = None


# this line gets sets the parameters for the HTML API
rest = GerritRestAPI(url=gerrit_url, auth = auth)


logger.info('REST API set-up complete')
```

```
# a set of parameters for the JSON API to get changes in batches of
500
start = 0                            # which batch we start from - usually
0

logger.info('Connecting to Gerrit server and accessing changes')

try:
    # the main query where we ask the endpoing to provide us the list
and details of all changes
    # each change is essentially a review that has been submitted to
the repository
    changes = rest.get("/changes/?q=status:merged&o=ALL_FILES&o=ALL_
REVISIONS&o=DETAILED_LABELS&start={}".format(start),
                        headers={'Content-Type': 'application/json'})
except Exception as e:
    logger.error('ENTITY ACCESS - Error retrieving changes: {}'.
format)

logger.info('Changes retrieved')
```

The boldface lines in the preceding code fragment log the information messages with the `logger.info(...)` statement as well as error messages with the `logger.error(...)` statement.

The content of this log file can be quite substantial, so we need to filter the messages based on their importance. That's why we distinguish between errors and information.

The following is a fragment of such a log file. The first line contains the information message (boldface INFO) to show that the machine learning pipeline has been started:

```
2023-01-15 17:11:45,618;Gerrit data export pipeline;INFO;Configuration
started
2023-01-15 17:11:45,951;Gerrit data export pipeline;INFO;Configuration
ended
2023-01-15 17:11:46,052;Gerrit data export pipeline;INFO;Downloading
fresh data to ./gerrit_reviews.csv
2023-01-15 17:11:46,055;pygerrit2;DEBUG;Error parsing netrc: netrc
missing or no credentials found in netrc
2023-01-15 17:11:46,057;Gerrit data export pipeline;INFO;Geting the
data about changes from 0 to 500
2023-01-15 17:11:46,060;urllib3.connectionpool;DEBUG;Starting new
HTTPS connection (1): gerrit.onap.org:443
```

We filter these messages in the last part of our measurement system – the information quality measurement and monitoring system. This last part reads through the log files and collects the error messages, categorizes them, and then visualizes them:

```python
try:
    logFile = open("./information_quality_gerrit.log", "r")

    for logMessage in logFile:
        # splitting the log information - again, this is linked to how we
        # structured the log message in the measurement system
        logItem = logMessage.split(';')
        logLevel = logItem[2]
        logSource = logItem[1]
        logTime = logItem[0]
        logProblem = logItem[3]

        # this part is about extracting the relevant information
        # if this is a problem at all:
        if (logLevel == 'ERROR'):
            # if this is a problem with the library
            if ('LIBRARIES' in logProblem):
                iq_problems_configuration_libraries += 1
            if ('ENTITY_ACCESS' in logProblem):
                iq_problems_entity_access += 1

except Exception as e:
    iq_general_error = 1
```

The bold-faced lines categorize the error messages found – in other words, they quantify the quality dimensions. This quantification is important as we need to understand how many problems of each kind were found in the machine learning pipeline.

The next step is to visualize the information quality, and for that, we need a quality analysis model. Then, we can use this quality model to visualize the quality dimensions:

```python
def getIndicatorColor(ind_value):
    if ind_value > 0:
        return 'red'
    else:
        return 'green'

columns = ('Information quality check', 'Value')
rows = ['Entity access check', 'Libraries']
cell_text = [[f'Entity access: {iq_problems_entity_access}'],
```

```
                  [f'Libraries: {iq_problems_configuration_libraries}']]
colors = [[getIndicatorColor(iq_problems_entity_access)],
                [getIndicatorColor(iq_problems_configuration_libraries)]]
```

The visualization can be done in several ways, but in the majority of cases, it is enough to visualize it in a tabular form, which is easy to overview and comprehend. The most important for this visualization is that it communicates whether there are (or not) any information quality problems:

```
fig, ax = plt.subplots()
ax.axis('tight')
ax.axis('off')
the_table = ax.table(cellText=cell_text,
                     cellColours=colors,
                     colLabels=columns,
                     loc='left')

plt.show()
```

The result of this code fragment is the visual representation shown in *Figure 4.5*. This example can be found in this book's GitHub repository:

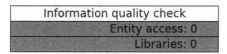

Figure 4.5 – Results from the quality checks, visualized in a tabular form

This rudimentary way of checking the information's quality illustrates my next best practice.

> **Best practice # 28**
>
> Use simple logging to trace any problems in your machine learning pipeline to monitor information quality.

It's generally a good practice to design and develop robust software systems. Logging is one of the mechanisms that's used to detect potential problems. Logging is also a very good software engineering practice for systems that are not interactive, such as machine learning-based ones. Therefore, extracting the information from logs can help us understand the quality of the information that is used in a machine learning-based system.

Noise

Data quality in machine learning systems has one additional and crucial attribute – noise. Noise can be defined as data points that contribute negatively to the ability of machine learning systems to identify patterns in the data. These data points can be outliers that make the datasets skew toward one or several classes in classification problems. The outliers can also cause prediction systems to over- or under-predict because they emphasize patterns that do not exist in the data.

Another type of noise is contradictory entries, where two (or more) identical data points are labeled with different labels. We can illustrate this with the example of product reviews on Amazon, which we saw in *Chapter 3*. Let's import them into a new Python script with `dfData = pd.read_csv('./book_chapter_4_embedded_1k_reviews.csv')`. In this case, this dataset contains a summary of the reviews and the score. We focus on these two columns and we define noise as different scores for the same summary review. For example, if one person provides a score of 5 for the review with the tag "Awesome!" and another person provides a score of 4 for another review with the tag "Awesome!," the same data point becomes noisy as it is annotated with two different labels – two different scores.

So, first, we must check whether there are any duplicate entries:

```
# now, let's check if there are any duplicate entries

# get the number of all data points
allDataPoints = len(dfData.Summary)

# and get the number of unique data points
uniqueDataPoints = len(dfData.Summary.unique())

# check if the number of unique and all data points is the same
if allDataPoints != uniqueDataPoints:
  print(f'There are {allDataPoints - uniqueDataPoints} duplicate
entries, which can potentially be noisy')
```

This code checks whether the number of data points is the same as the number of unique data points; if not, then we risk having noisy entries. We can check whether there are duplicate entries by using the following code:

```
# then, we find duplicate data points

# first we group the datapoints
dfGrouped = dfData.groupby(by=dfData.Summary).count()

# then we find the index of the ones that are not unique
lstDuplicated = dfGrouped[dfGrouped.Time > 1].index.to_list()
```

Now, we can remove all duplicate entries using the following command, though a simple solution would be to remove them (`dfClean = dfData[~dfData.Summary.isin(lstDuplicated)]`). A better solution is to check whether they are noisy entries or just duplicates. We can do this using the following code fragment:

```
# for each of these data points, we check if these data points
# are classified to different labels adn remove only the ones that
have different labels
for onePoint in lstDuplicated:
    # find all instances of this datapoint
    dfPoint = dfData[dfData.Summary == onePoint]

    # now check if these data points have a different score
    numLabels = len(dfPoint.Score.unique())

    # if the number of labels is more than 1, then
    # this means that we have noise in the dataset
    # and we should remove this point
    if numLabels > 1:
        dfData.drop(dfData[dfData.Summary == onePoint].index,
inplace=True)

        # let's also print the data point that we remove
        print(f'point: {onePoint}, number of labels: {len(dfPoint.Score.
unique())}')
```

After running this fragment of code, the dataset does not contain any contradictory entries and therefore no class noise. Although it is possible to adopt a different strategy – for example, instead of removing noisy data points, we could change them to one of the classes – such an approach changes the pattern in the data, and therefore is not fully representative of the data. We simply do not know which of the classes is more correct than the others, especially if there are duplicate data points with two different labels.

Best practice # 29

The best strategy to reduce the impact of noise on machine learning classifiers is to remove the noisy data points.

Although we can correct noisy data points by changing their label or reducing the impact of these attributes on the predictions, the best strategy is to remove these data points. Removing is better as it does not change the patterns in the data. Imagine that we relabel noisy entries – this creates a pattern in the data that does not exist, which causes the algorithms to mispredict future data points.

Removing noise from the data is the only one way to handle noise. Another method is to increase the number of features so that we can distinguish between data points. We can analyze data and identify whether there is a risk of noise, and then we can check whether it is possible to add one more feature to the dataset to distinguish between entries labeled differently. However, this is outside the scope of this chapter.

Summary

Data for machine learning systems is crucial – without data, there can be no machine learning systems. In most machine learning literature, the process of training models usually starts with the data in tabular form. In software engineering, however, this is an intermediate step. The data is collected from source systems and needs to be processed.

In this chapter, we learned how to access data from modern software engineering systems such as Gerrit, GitHub, JIRA, and Git. The code included in this chapter illustrates how to collect data that can be used for further steps in the machine learning pipeline – feature extraction. We'll focus on this in the next chapter.

Collecting data is not the only preprocessing step that is required to design and develop a reliable software system. Quantifying and monitoring information (and data) quality is equally important. We need to check that the data is fresh (timely) and that there are no problems in preprocessing that data.

One of the aspects that is specific to machine learning systems is the presence of noise in the data. In this chapter, we learned how to treat class noise in the data and how to reduce the impact of the noise on the final machine learning algorithm.

In the next chapter, we dive deeper into concepts related to data - clearning it from noise and quantifying its properties.

References

- *Vaswani, A. et al., Attention is all you need. Advances in neural information processing systems, 2017. 30.*

- *Dastin, J., Amazon scraps secret AI recruiting tool that showed bias against women. In Ethics of Data and Analytics. 2018, Auerbach Publications. p. 296-299.*

- *Staron, M., D. Durisic, and R. Rana, Improving measurement certainty by using calibration to find systematic measurement error—a case of lines-of-code measure. In Software Engineering: Challenges and Solutions. 2017, Springer. p. 119-132.*

- *Staron, M. and W. Meding, Software Development Measurement Programs. Springer. https://doi.org/10.1007/978-3-319-91836-5, 2018. 10: p. 3281333.*

- *Fenton, N. and J. Bieman, Software metrics: a rigorous and practical approach. 2014: CRC press.*

- Li, N., M. Shepperd, and Y. Guo, *A systematic review of unsupervised learning techniques for software defect prediction. Information and Software Technology*, 2020. 122: p. 106287.

- Staron, M. et al. *Robust Machine Learning in Critical Care—Software Engineering and Medical Perspectives. In 2021 IEEE/ACM 1st Workshop on AI Engineering-Software Engineering for AI (WAIN).* 2021. IEEE.

- Zhang, J. et al., *CoditT5: Pretraining for Source Code and Natural Language Editing. arXiv preprint arXiv:2208.05446*, 2022.

- Staron, M. et al. *Using machine learning to identify code fragments for manual review. In 2020 46th Euromicro Conference on Software Engineering and Advanced Applications (SEAA).* 2020. IEEE.

- Ochodek, M., S. Kopczyńska, and M. Staron, *Deep learning model for end-to-end approximation of COSMIC functional size based on use-case names. Information and Software Technology*, 2020. 123: p. 106310.

- Cichy, C. and S. Rass, *An overview of data quality frameworks. IEEE Access*, 2019. 7: p. 24634-24648.

- Lee, Y.W. et al., *AIMQ: a methodology for information quality assessment. Information & management*, 2002. 40(2): p. 133-146.

- Staron, M. and W. Meding. *Ensuring reliability of information provided by measurement systems. In International Workshop on Software Measurement.* 2009. Springer.

- Pandazo, K. et al. *Presenting software metrics indicators: a case study. In Proceedings of the 20th international conference on Software Product and Process Measurement (MENSURA).* 2010.

- Staron, M. et al. *Improving Quality of Code Review Datasets–Token-Based Feature Extraction Method. In International Conference on Software Quality.* 2021. Springer.

5
Quantifying and Improving Data Properties

Procuring data in machine learning systems is a long process. So far, we have focused on data collection from source systems and cleaning noise from data. Noise, however, is not the only problem that we can encounter in data. Missing values or random attributes are examples of data properties that can cause problems with machine learning systems. Even the length of the input data can be problematic if it is outside of the expected values.

In this chapter, we will dive deeper into the properties of data and how to improve them. In contrast to the previous chapter, we will work on feature vectors rather than raw data. Feature vectors are already a transformation of the data and therefore, we can change properties such as noise or even change how the data is perceived.

We'll focus on the processing of text, which is an important part of many machine learning algorithms nowadays. We'll start by understanding how to transform data into feature vectors using simple algorithms such as bag of words. We will also learn about techniques to handle problems in data.

In this chapter, we will cover the following main topics:

- Quantifying data properties for machine learning systems
- Germinating noise – feature engineering in clean datasets
- Handling noisy data – machine learning algorithms and noise removal
- Eliminating attribute noise – a guide to dataset refinement

Feature engineering – the basics

Feature engineering is the process of transforming raw data into vectors of numbers that can be used in machine learning algorithms. This process is structured and requires us to first select which feature extraction mechanism we need to use – which depends on the type of the task – and then configure the chosen feature extraction mechanism. When the chosen algorithm is configured, we can use it to transform the raw input data into a matrix of features – we call this process feature extraction. Sometimes, the data needs to be processed before (or after) the feature extraction, for example, by merging fields or removing noise. This process is called data wrangling.

The number of feature extraction mechanisms is large, and we cannot cover all of them. Not that we need to either. What we need to understand, however, is how the choice of feature extraction mechanism influences the properties of the data. We'll dive much deeper into the process of feature engineering in the next chapter, but in this chapter, we will introduce a basic algorithm for textual data. We need to introduce it to understand how it impacts the properties of data and how to cope with the most common problems that can arise in the context of feature extraction, including dealing with "dirty" data that requires cleaning.

To understand this process, let us start with the first example of feature extraction from text using an algorithm called bag of words. Bag of words is a method that transforms a piece of text into a vector of numbers showing which words are part of that text. The words form the set of features – or columns – in the resulting dataframe. In the following code, we can see how feature extraction works. We have used the standard library from `sklearn` to create a bag-of-words feature vector.

In the following code fragment, we take two lines of C code – `printf("Hello world!");` and `return 1` – and then translate this into a matrix of features:

```
# create the feature extractor, i.e., BOW vectorizer
# please note the argument - max_features
# this argument says that we only want three features
# this will illustrate that we can get problems - e.g. noise
# when using too few features
vectorizer = CountVectorizer(max_features = 3)
# simple input data - two sentences
sentence1 = 'printf("Hello world!");'
sentence2 = 'return 1'
# creating the feature vectors for the input data
X = vectorizer.fit_transform([sentence1, sentence2])

# creating the data frame based on the vectorized data
df_bow_sklearn = pd.DataFrame(X.toarray(),
                              columns=vectorizer.get_feature_names(),
                              index=[sentence1, sentence2])

# take a peek at the featurized data
df_bow_sklearn.head()
```

The line in bold is a statement that creates an instance of the `CodeVectorizer` class, which transforms a given text into a vector of features. This includes the extraction of the features identified.

This line has one parameter – `max_features` `= ` 3. This parameter tells the algorithm that we only want three features. In this algorithm, the features are the words that are used in the input text. When we input the text to the algorithm, it extracts the tokens (words), and then for every line, it counts whether it contains these words. This is done in the statement `X = vectorizer.fit_transform([sentence1, sentence2])`. When the features are extracted, the resulting dataset looks as follows:

	Hello	printf	return
printf("Hello world!");	1	1	0
return 1	0	0	1

Figure 5.1 – Extracted features create this dataset

The first line in the table contains the index – the line that was input to the algorithm – and then `1` or `0` to show that the line contains the words in the vocabulary. Since we only asked for three features, the table has three columns – `Hello`, `printf`, and `return`. If we change the parameter of the `CountVectorizer()`, we'll obtain the full list of tokens in these two lines, that is, `hello`, `printf`, `return`, and `world`.

For these two simple lines of C code, we get four features, which illustrates that this kind of feature extraction can quickly increase the size of the data. This leads us on to my next best practice.

> **Best practice #30**
>
> Balance the number of features with the number of data points. More features is not always better.

When creating feature vectors, it is important to extract meaningful features that can effectively distinguish between the data points. However, we should keep in mind that having more features will require more memory and can make the training slower. It is also prone to problems with missing data points.

Clean data

One of the most problematic aspects of datasets, when it comes to machine learning, is the presence of empty data points or empty values of features for data points. Let's illustrate that with the example of the features extracted in the previous section. In the following table, I introduced an empty data point – the NaN value in the middle column. This means that the value does not exist.

	Hello	printf	return
printf("Hello world!");	1	NaN	0
return 1	0	0	1

Figure 5.2 – Extracted features with a NaN value in the table

If we use this data as input to a machine learning algorithm, we'll get an error message that the data contains empty values and that the model cannot be trained. That is a very accurate description of this problem – if there is a missing value, then the model does not know how to handle it and therefore it cannot be trained.

There are two strategies to cope with empty values in datasets – removing the data points or imputing the values.

Let's start with the first strategy – removing the empty data points. The following script reads the data from our code reviews that we will use for further calculations:

```
# read the file with gerrit code reviews
dfReviews = pd.read_csv('./gerrit_reviews.csv', sep=';')


# just checking that we have the right columns
# and the right data
dfReviews.head()
```

The preceding fragment of code reads the file and displays its first 10 rows for us to inspect what the data contains.

Once we have the data in memory, we can check how many of the rows contain null values for the column that contains the actual line of code, which is named LOC. Then, we can also remove the rows/data points that do not contain any data. The removal of the data points is handled by the following line – dfReviews = dfReviews.dropna(). This statement removes the lines that are empty and keeps the result in the dataframe itself (the inplace=True parameter):

```
import numpy as np
# before we use the feature extractor, let's check if the data
contains NANs
print(f'The data contains {dfReviews.LOC.isnull().sum()} empty rows')


# remove the empty rows
dfReviews = dfReviews.dropna()


# checking again, to make sure that it does not contain them
print(f'The data contains {dfReviews.LOC.isnull().sum()} empty rows')
```

After these commands, our dataset is prepared to create the feature vector. We can use CountVectorizer to extract the features from the dataset, as in the following code fragment:

```
# now, let's convert the code (LOC) column to the vector of features
# using BOW from the example above
vectorizer = CountVectorizer(min_df=2,
                             max_df=10)


dfFeatures = vectorizer.fit_transform(dfReviews.LOC)
```

```
# creating the data frame based on the vectorized data
df_bow_sklearn = pd.DataFrame(dfFeatures.toarray(),
                                 columns=vectorizer.get_feature_
names(),index=dfReviews.LOC)

# take a peek at the featurized data
df_bow_sklearn.head()
```

This fragment creates the bag-of-words model (CountVectorizer) with two parameters – the minimum frequency of the tokens and the maximum frequency of the tokens. This means that the algorithm calculates the statistics of how frequently each token appears in the dataset and then chooses the ones that fulfill the criteria. In our case, the algorithm chooses the tokens that appear at least twice (min_df=2) and at most 20 times (max_df=20).

The result of this code fragment is a large dataframe with 661 features extracted for each line of code in our dataset. We can check this by writing len(df_bow_sklearn.columns) after executing the preceding code fragment.

In order to check how to work with data imputation, let us open a different dataset and check how many missing data points we have per column. Let's read the dataset that is named gerrit_reviews_nan.csv and list the number of missing values in that dataset using the following code fragment:

```
# read data with NaNs to a dataframe
dfNaNs = pd.read_csv('./gerrit_reviews_nan.csv', sep='$')

# before we use the feature extractor, let's check if the data
contains NANs
print(f'The data contains {dfNaNs.isnull().sum()} NaN values')
```

As a result of this code fragment, we get a list of columns with the number of missing values in them – the tail of this list is as follows:

```
yangresourcesnametocontentmap        213
yangtextschemasourceset              205
yangtextschemasourcesetbuilder       208
yangtextschemasourcesetcache         207
yangutils                            185
```

There are many missing values and therefore, we need to adopt another strategy than removing them. If we remove all these values, we get exactly 0 data points – which means that there is a NaN value in one (or more) of the columns for every(!) data point. So, we need to adopt another strategy – imputation.

First, we need to prepare the data for the imputer, which only works on the features. Therefore, we need to remove the index from the dataset:

```
# in order to use the imputer, we need to remove the index from the
data
# we remove the index by first re-setting it (so that it becomes a
regular column)
# and then by removing this column.
dfNaNs_features = dfNaNs.reset_index()
dfNaNs_features.drop(['LOC', 'index'], axis=1, inplace=True)
dfNaNs_features.head()
```

Then, we can create the imputer. In this example, I use one of the modern ones, which is based on training a classifier on the existing data and then using it to fill the data in the original dataset. The fragment of code that trains the imputer is presented here:

```
# let's use iterative imputed to impute data to the dataframe
from sklearn.experimental import enable_iterative_imputer
from sklearn.impute import IterativeImputer

# create the instance of the imputer
imp = IterativeImputer(max_iter=3,
                       random_state=42,
                       verbose = 2)

# train the imputer on the features in the dataset
imp.fit(dfNaNs_features)
```

The last line of the code fragment is the actual training of the imputer. After this, we can start making the imputations to the dataset, as in the following code fragment:

```
# now, we fill in the NaNs in the original dataset
npNoNaNs = imp.transform(dfNaNs_features)
dfNoNaNs = pd.DataFrame(npNoNaNs)
```

After this fragment, we have a dataset that contains imputer values. Now, we need to remember that these values are only estimations, not the real ones. This particular dataset illustrates this very well. When we execute the dfNoNaNs.head() command, we can see that some of the imputed values are negative. Since our dataset is the result of CountVectorizer, the negative values are not likely. Therefore, we could use another kind of imputer – KNNImputer. That imputer uses the nearest neighbor algorithm to find the most similar data points and fills in the missing data based on the values of the similar data points. In this way, we get a set of imputed values that have the same properties (e.g., no negative values) as the rest of the dataset. However, the pattern of the imputed values is different.

Therefore, here is my next best practice.

> **Best practice #30**
> Use KNNImputer for data where the similarity between data points is expected to be local.

`KNNImputer` works well when there is a clear local structure in the data, especially when neighboring data points are similar in terms of the feature with missing values. It can be sensitive to the choice of the number of nearest neighbors (`k`).

`IterativeImputer` tends to perform well when there are complex relationships and dependencies among features in the dataset. It may be more suitable for datasets with missing values that are not easily explained by local patterns.

However, check whether the imputation method provides logical results for the dataset at hand, in order to reduce the risk of bias.

Noise in data management

Missing data and contradictory annotations are only one type of problem with data. In many cases, large datasets, which are generated by feature extraction algorithms, can contain too much information. Features can be superfluous and not contribute to the end results of the algorithm. Many machine learning models can deal with noise in the features, called attribute noise, but too many features can be costly in terms of training time, storage, and even data collection itself.

Therefore, we should also pay attention to the attribute noise, identify it, and then remove it.

Attribute noise

There are a few methods to reduce attribute noise in large datasets. One of these methods is an algorithm named the **Pairwise Attribute Noise Detection Algorithm** (**PANDA**). PANDA compares features pairwise and identifies which of them adds noise to the dataset. It is a very effective algorithm, but unfortunately very computationally heavy. If our dataset had a few hundred features (which is when we would really need to use this algorithm), we would need a lot of computational power to identify these features that bring in little to the analysis.

Fortunately, there are machine learning algorithms that provide similar functionality with little computational overhead. One of these algorithms is the random forest algorithm, which allows you to retrieve the set of feature importance values. These values are a way of identifying which features are not used in any of the decision trees in this forest.

Let us then see how to use that algorithm to extract and visualize the feature's importance. For this example, we will use the data extracted from the Gerrit tool in previous chapters:

```python
# importing the libraries to vectorize text
# and to manipulate dataframes
from sklearn.feature_extraction.text import CountVectorizer
import pandas as pd

# create the feature extractor, i.e., BOW vectorizer
# please note the argument - max_features
# this argument says that we only want three features
# this will illustrate that we can get problems - e.g. noise
# when using too few features
vectorizer = CountVectorizer()

# read the file with gerrit code reviews
dfReviews = pd.read_csv('./gerrit_reviews.csv', sep=';')
```

In this dataset, we have two columns that we extract features from. The first is the LOC column, which we use to extract the features using CountVectorizer – just like in the previous example. These features will be our X values later for the training algorithm. The second column of interest is the message column. The message column is used to provide the decision class. In order to transform the text of the message, we use a sentiment analysis model to identify whether the message is positive or negative.

First, let's extract the BOW features using CountVectorizer:

```python
# now, let's convert the code (LOC) column to the vector of features
# using BOW from the example above
vectorizer = CountVectorizer(min_df=2,
                             max_df=10)

dfFeatures = vectorizer.fit_transform(dfReviews.LOC)

# creating the data frame based on the vectorized data
df_bow_sklearn = pd.DataFrame(dfFeatures.toarray(),
                              columns=vectorizer.get_feature_
names(),index=dfReviews.LOC)
```

To transform the message into a sentiment, we can use an openly available model from the Hugging Face Hub. We need to install the relevant libraries using the following command: ! pip install -q transformers. Once we have the libraries, we can start the feature extraction:

```python
# using a classifier from the Hugging Face hub is quite
straightforward
```

```
# we import the package and create the sentiment analysis pipeline
from transformers import pipeline

# when we create the pipeline, and do not provide the model
# then the huggingface hub will choose one for us
# and download it
sentiment_pipeline = pipeline("sentiment-analysis")

# now we are ready to get the sentiment from our reviews.
# let's supply it to the sentiment analysis pipeline
lstSentiments = sentiment_pipeline(list(dfReviewComments))

# transform the list to a dataframe
dfSentiments = pd.DataFrame(lstSentiments)

# and then we change the textual value of the sentiment to
# a numeric one - which we will use for the random forest
dfSentiment = dfSentiments.label.map({'NEGATIVE': 0, 'POSITIVE': 1})
```

The preceding code fragment uses the pre-trained model for the sentiment analysis and one from the standard pipeline – sentiment-analysis. The result is a dataframe that contains a positive or negative sentiment.

Now, we have both the X values – features extracted from the lines of code – and the predicted Y values – the sentiment from the review comment message. We can use these to create a dataframe that we can use as an input to the random forest algorithm, train the algorithm, and identify which features contributed the most to the result:

```
# now, we train the RandomForest classifier to get the most important
features
# Note! This training does not use any data split, as we only want to
find
# which features are important.
X = df_bow_sklearn.drop(['sentiment'], axis=1)
Y = df_bow_sklearn['sentiment']

# import the classifier - Random Forest
from sklearn.ensemble import RandomForestClassifier

# create the classifier
clf = RandomForestClassifier(max_depth=10, random_state=42)

# train the classifier
# please note that we do not check how good the classifier is
```

```
# only train it to find the features that are important.
Clf.fit(X,Y)
```

When the random forest model is trained, we can extract the list of important features:

```
# now, let's check which of the features are the most important ones
# first we create a dataframe from this list
# then we sort it descending
# and then filter the ones that are not important
dfImportantFeatures = pd.DataFrame(clf.feature_importances_, index=X.
columns, columns=['importance'])

# sorting values according to their importance
dfImportantFeatures.sort_values(by=['importance'],
                                ascending=False,
                                inplace=True)

# choosing only the ones that are important, skipping
# the features which have importance of 0
dfOnlyImportant =
dfImportantFeatures[dfImportantFeatures['importance'] != 0]

# print the results
print(f'All features: {dfImportantFeatures.shape[0]}, but only
{dfOnlyImportant.shape[0]} are used in predictions. ')
```

This preceding code fragment selects the features with an importance of more than 0 and then lists them. We find that 363 out of 662 features are used in the predictions. This means that the remaining 270 are just the attribute noise.

We can also visualize these features using the seaborn library, as in the following code fragment:

```
# we use matplotlib and seaborn to make the plot
import matplotlib.pyplot as plt
import seaborn as sns

# Define size of bar plot
# We make the x axis quite much larger than the y-axis since
# there is a lot of features to visualize
plt.figure(figsize=(40,10))

# plot seaborn bar chart
# we just use the blue color
sns.barplot(y=dfOnlyImportant['importance'],
            x=dfOnlyImportant.index,
```

```
              color='steelblue')

# we make the x-labels rotated so that we can fit
# all the features
plt.xticks(rotation=90)

# add chart labels
plt.title('Importance of features, in descending order')
plt.xlabel('Feature importance')
plt.ylabel('Feature names')
```

This preceding code fragment results in the following diagram for the dataset:

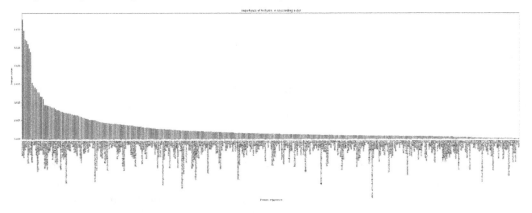

Figure 5.3 – Feature importance chart with numerous features

Since there are so many features, the diagram gets very cluttered and challenging to read, so we can only visualize the top 20 features to understand which ones are really the most important.

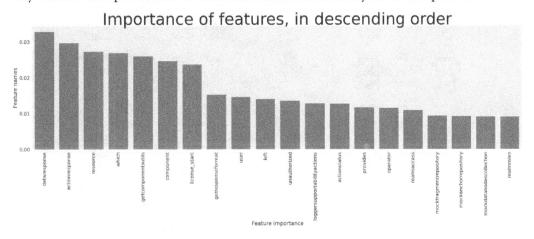

Figure 5.4 – Top 20 most important features in the dataset

The preceding code examples show that we can reduce the number of features by 41%, which is almost half of the features. The algorithm takes just a few seconds to find the most important features, which makes it the perfect candidate for reducing attribute noise in the datasets.

> **Best practice #31**
> Use the Random Forest classifier to eliminate unnecessary features, as it offers very good performance.

Although we do not really get information on how much noise the removed features contain, receiving information that they have no value for the prediction algorithm is sufficient. Therefore, I recommend using this kind of feature reduction technique in the machine learning pipeline in order to reduce the computational and storage needs of our pipeline.

Splitting data

For the process of designing machine learning-based software, another important property is to understand the distribution of data, and, subsequently, ensure that the data used for training and testing is of a similar distribution.

The distribution of the data used for training and validation is important as the machine learning models identify patterns and re-create them. This means that if the data in the training is not distributed in the same way as the data in the test set, our model misclassifies data points. The misclassifications (or mispredictions) are caused by the fact that the model learns patterns in the training data that are different from the test data.

Let us understand how splitting algorithms work in theory, and how they work in practice. *Figure 5.5* shows how the splitting works on a theoretical and conceptual level:

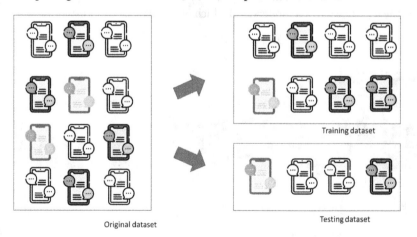

Original dataset

Training dataset

Testing dataset

Figure 5.5 – Splitting data into train and test sets

Icons represent review comments (and discussions). Every icon symbolizes its own discussion thread, and each type of icon reflects different teams. The idea behind splitting the dataset is that the two sets are very similar, but not identical. Therefore, the distribution of elements in the training and test datasets needs to be as similar as possible. However, it is not always possible, as *Figure 5.5* shows – there are three out of four icons of one of the kinds in the training set and only one in the test set. When designing machine learning software, we need to take this aspect into consideration, even though it only relates to machine learning models. Our data processing pipeline should contain checks that provide the ability to understand whether the data is correctly distributed and, if not, we need to correct it. If we do not correct it, our system starts mispredicting. The change in the distribution of data over time, which is natural in machine learning-based systems, is called concept drift.

Let us use this in practice by calculating the distributions of the data in our Gerrit reviews dataset. First, we read the data, and then we use the sklearn `train_test_split` method to create a random split:

```python
# then we read the dataset
dfData = pd.read_csv('./bow_sentiment.csv', sep='$')


# now, let's split the data into train and test
# using the random split
from sklearn.model_selection import train_test_split


X = dfData.drop(['LOC', 'sentiment'], axis=1)
y = dfData.sentiment


# now we are ready to split the data
# test_size parameter says that we want 1/3rd of the data in the test
set
# random state allows us to replicate the same split over and over
again
X_train, X_test, y_train, y_test =
                train_test_split(X, y,
                                test_size=0.33,
                                random_state=42)
```

In this code fragment, we separate the predicted values (y) from the predictor values (X) features. Then we use the `train_test_split` method to split the dataset into two – two-thirds of the data in the training set and one-third of the data in the test set. This 2:1 ratio is the most common, but we can also encounter a 4:1 ratio, depending on the application and the dataset.

Now that we have two sets of data, we should explore whether the distributions are similar. Essentially, we should do that for each feature and the predicted variable (y), but in our dataset, we have 662 features, which means that we would have to do as many comparisons. So, let us, for the sake of the example, visualize only one of them – the one that was deemed the most important in our previous example – dataresponse:

```
# import plotting libraries
import matplotlib.pyplot as plt
import seaborn as sns


# we make the figure a bit larger
# and the font a bit more visible
plt.figure(figsize=(10,7))
sns.set(font_scale=1.5)


# here we visualize the histogram using seaborn
# we take only one of the variables, please see the list of columns
# above, or use print(X_train.columns) to get the list
# I chose the one that was the most important one
# for the prediction algorithm
sns.histplot(data=X_train['dataresponse'],
            binwidth=0.2)
```

We will do the same for the test set too:

```
plt.figure(figsize=(10,7))
sns.set(font_scale=1.5)


sns.histplot(data=X_test['dataresponse'],
            binwidth=0.2)
```

These two fragments result in two histograms with the distribution for that variable. They are presented in *Figure 5.6*:

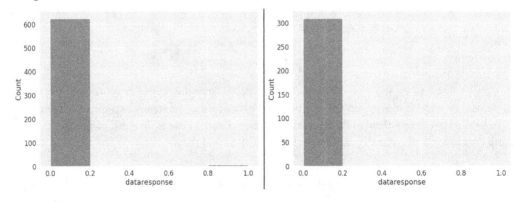

Figure 5.6 – Distribution of dataresponse feature in train and test set

The train set's distribution is on the left-hand side and the test set's distribution is on the right-hand side. At first glance, the distributions show that there is only a single value - 0 value. Therefore, we need to explore the data manually a bit more. We can check the distribution by calculating the number of entities per value – 0 and 1:

```
# we can even check the count of each of these values
X_train_one_feature = X_train.groupby(by='dataresponse').count()
X_train_one_feature

# we can even check the count of each of these values
X_test_one_feature = X_test.groupby(by='dataresponse').count()
X_test_one_feature
```

From the preceding calculations, we find that there are 624 values of 0 and 5 values of 1 in the train set. We also find that there are 309 values of 0 and 1 value of 1 in the test set. These are not exactly the same ratio, but given the scale – the 0s are significantly more than the 1s – this does not have any impact on the machine learning model.

The features in our dataset should have the same distribution, but so do the Y values – the predicted variables. We can use the same technique to visualize the distribution between classes in the Y value. The following code fragment does just that:

```
# we make the figure a bit larger
# and the font a bit more visible
plt.figure(figsize=(10,7))
sns.set(font_scale=1.5)

sns.histplot(data=y_train, binwidth=0.5)
sns.histplot(data=y_test,  binwidth=0.5)
```

This code fragment generates two diagrams, which show what the difference between the two classes is. They are presented in *Figure 5.7*:

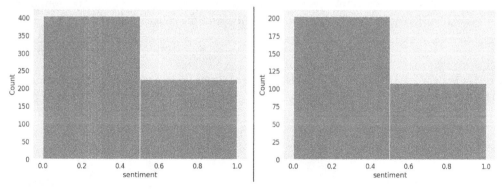

Figure 5.7 – Distribution of classes (0 and 1) in the training and test data

The predicted Y variable 0s are the negative sentiment values while 1s are the positive ones. Although the scales on the *y* axis are different in both diagrams, the distributions are very similar – it is roughly 2:1 in terms of the number of negative (0) sentiments and positive (1) sentiments.

The classes are not balanced – the number of 0s is much larger than the number of 1s, but the distribution is the same. The fact that the classes are not balanced means that the model trained on this data is slightly biased towards the negative sentiment rather than the positive sentiment. However, this reflects the empirical observations that we make: in code reviews, the reviewers are more likely to comment on code that needs to be improved rather than on code that is nicely written.

> **Best practice #32**
>
> As much as possible, retain the original distribution of the data as it reflects the empirical observations.

Although we can balance the classes using undersampling, oversampling, or similar techniques, we should consider keeping the original distribution as much as we can. Changing the distribution makes the model "fairer" in terms of predictions/classifications, but it changes the nature of the observed phenomena.

How ML models handle noise

Reducing noise from datasets is a time-consuming task, and it is also a task that cannot be easily automated. We need to understand whether we have noise in the data, what kind of noise is in the data, and how to remove it. Luckily, most machine learning algorithms are pretty good at handling noise.

For example, the algorithm that we have used quite a lot so far – random forest – is quite robust to noise in datasets. Random forest is an ensemble model, which means that it is composed of several separate decision trees that internally "vote" for the best result. This voting process can therefore filter out noise and coalescence toward the pattern contained in the data.

Deep learning algorithms have similar properties too – by utilizing a number of small neurons, these networks are robust to noise in large datasets. They can coerce the pattern in the data.

> **Best practice #33**
>
> In large-scale software systems, if possible, rely on machine learning models to handle noise in the data.

It may sound like I'm proposing an easy way out, but I'm not. Manual cleaning of the data is crucial, but it is also slow and costly. Therefore, during operations in large-scale systems, it is better to select a model that is robust to noise in the data and at the same time uses cleaner data. Since manual noise-handling processes require time and effort, relying on them would introduce unnecessary costs for our product operations.

Therefore, it's better to use algorithms that do that for us and therefore create products that are reliable and require minimal maintenance. Instead of costly noise-cleaning processes, it's much more cost-efficient to re-train the algorithm to let it do the work for you.

In the next chapter, we explore data visualization techniques. These techniques help us to understand dependencies in the data and whether it exposes characteristics that can be learnt by the machine learning models.

References

- *Scott, S. and S. Matwin. Feature engineering for text classification. in ICML. 1999.*

- *Kulkarni, A., et al., Converting text to features. Natural Language Processing Recipes: Unlocking Text Data with Machine Learning and Deep Learning Using Python, 2021: p. 63-106.*

- *Van Hulse, J.D., T.M. Khoshgoftaar, and H. Huang, The pairwise attribute noise detection algorithm. Knowledge and Information Systems, 2007. 11: p. 171-190.*

- *Li, X., et al., Exploiting BERT for end-to-end aspect-based sentiment analysis. arXiv preprint arXiv:1910.00883, 2019.*

- *Xu, Y. and R. Goodacre, On splitting training and validation set: a comparative study of cross-validation, bootstrap and systematic sampling for estimating the generalization performance of supervised learning. Journal of analysis and testing, 2018. 2(3): p. 249-262.*

- *Mosin, V., et al. Comparing Input Prioritization Techniques for Testing Deep Learning Algorithms. in 2022 48th Euromicro Conference on Software Engineering and Advanced Applications (SEAA). 2022. IEEE.*

- *Liu, X.-Y., J. Wu, and Z.-H. Zhou, Exploratory undersampling for class-imbalance learning. IEEE Transactions on Systems, Man, and Cybernetics, Part B (Cybernetics), 2008. 39(2): p. 539-550.*

- *Atla, A., et al., Sensitivity of different machine learning algorithms to noise. Journal of Computing Sciences in Colleges, 2011. 26(5): p. 96-103.*

Part 2: Data Acquisition and Management

Machine learning software depends on data much more than other types of software. In order to make use of statistical learning, we need to collect, process, and prepare data for the development of machine learning models. The data needs to be representative of the problems that the software solves and the services it provides, not only during development but also during operations. In this part of the book, we focus on the data – how we can acquire it and how we make it useful for the training, testing, and deployment of machine learning models.

This part has the following chapters:

- *Chapter 6, Processing Data in Machine Learning Systems*
- *Chapter 7, Feature Engineering for Numerical and Image Data*
- *Chapter 8, Feature Engineering for Natural Language Data*

Processing Data in Machine Learning Systems

We talked about data in *Chapter 3*, where we introduced the types of data that are used in machine learning systems. In this chapter, we'll dive deeper into ways in which data and algorithms are entangled. We'll talk about data in generic terms, but in this chapter, we'll explain what kind of data is needed in machine learning systems. I'll explain the fact that all kinds of data are used in numerical form – either as a feature vector or as more complex feature matrices. Then, I'll explain the need to transform unstructured data (for example, text) into structured data. This chapter will lay the foundations for diving deeper into each type of data, which is the content of the next few chapters.

In this chapter, we will do the following:

- Discuss the process of measurement (obtaining numerical data) and the measurement instruments that are used in that process
- Visualize numerical data using the Matplotlib and Seaborn libraries
- Reduce dimensions using **principal component analysis (PCA)**
- Work with Hugging Face's Dataset module to download and process image and text data

Numerical data

Numerical data usually comes in the form of tables of numbers, kind of like database tables. One of the most common data in this form is metrics data – for example, the standard object-oriented metrics that have been used since the 1980s.

Numerical data is often the result of a measurement process. The measurement process is a process where we quantify the empirical properties of an entity using measurement instruments to a number. The process must guarantee that important empirical properties are preserved in the mathematical domain – that is, in the numbers. *Figure 6.1* shows an example of this process:

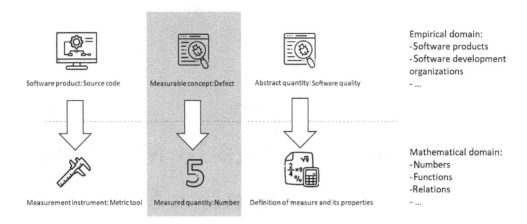

Figure 6.1 – The measurement process with an example of quality measurement using defects

The important part of this process consists of three elements. First is the measurement instrument, which needs to map the empirical properties to numbers in a true way. Then, there are the measurement standards, such as the ISO **Vocabulary in Metrology** (**VIM**), which is called the measurement's trueness. Finally, we have the result of the measurement process – the process of applying the measurement instrument to a specific measured entity – which results in a number, a quantification of the measured property. A single number, however, cannot characterize an entire software product or even a part of it, regardless of how true it is to the measured entity. Therefore, in practice, we use several measurement instruments to create a holistic view of the measured entity.

This is where numerical data comes in. Each measurement that characterizes the measured entity is stored in a database or a table – each entity becomes one row and each metric becomes one column. The more columns we have, the better the characteristics of the measured entity. However, at the same time, the more measures we collect, the higher the risk that they will be interconnected, correlated (positively and negatively), and that they will overlap. So, we need to work a bit with that data to get some orientation of it. So, first, we must visualize the data.

The data we'll use in this part of this chapter comes from a paper by Alhustain, Sultan (Predicting Relative Thresholds for Object Oriented Metrics." 2021 IEEE/ACM International Conference on Technical Debt (TechDebt). IEEE, 2021) and is available from Zenodo (`https://zenodo.org/records/4625975`), one of the most commonly used open data repositories in software engineering research. The data contains the values of measures for typical object-oriented metrics:

- **Coupling between objects** (**CB**): The number of references to other classes from the measured entity (class)

- **Direct class coupling** (**DCC**): The number of connections from this class to other classes (for example, associations)

- **ExportCoupling**: The number of outgoing connections from the class

- **ImportCoupling**: The number of incoming connections to the class

- **Number of methods** (**NOM**): The number of methods in the class
- **Weighted methods per class** (**WMC**): The number of methods in the class, weighted by their size
- **Defect count** (**defect**): The number of defects discovered for this class

The dataset describes several software projects from the Apache foundation – for example, the Ant tool. For each product, the measured entities are classes in the project.

So, let's start with the next best practice, which will lead us to the visualization.

> **Best practice #34**
>
> When working with numerical data, visualize it first, starting with the summary views of the data.

When I work with numerical data, I usually start by visualizing it. I start with some overview of the data and then work my way toward the details.

Summarizing the data

Summarizing the data can be done using tables and pivots, as well as charts. One of the charts that I usually start working with is the correlogram – it's a diagram that shows correlations between each variable/measure in the dataset.

So, let's read our data into the notebook and start visualizing it:

```
# read the file with data using openpyxl
import pandas as pd

# we read the data from the excel file,
# which is the defect data from the ant 1.3 system
dfDataAnt13 = pd.read_excel('./chapter_6_dataset_numerical.xlsx',
                            sheet_name='ant_1_3',
                            index_col=0)
```

Once the dataset is in memory, we can use Python's Seaborn library to visualize it using the correlogram. The following code does just that:

```
# now, let's visualize the data using correlograms
# for that, we use the seaborn library
import seaborn as sns
import matplotlib.pyplot as plt

# in seaborn, the correlogram is called
# pairplot
sns.pairplot(dfDataAnt13)
```

The result of this code fragment is the following correlogram:

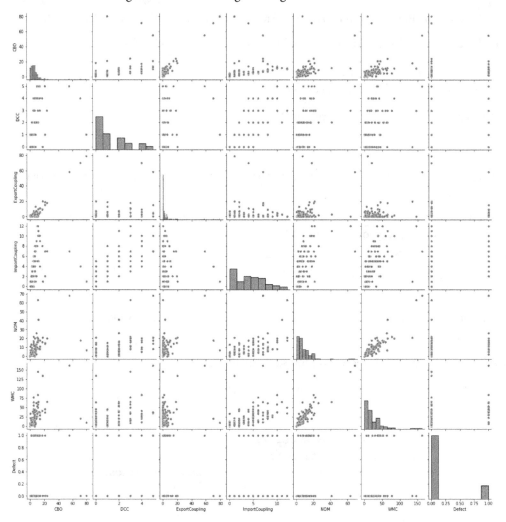

Figure 6.2 – Correlogram for the dataset from the paper of Alhusain

The interesting part here is the distribution of each of the measures presented in the cells on the diagonal. In the case of our data, this distribution is hard to interpret for some of the variables, so we can visualize it a bit differently. When we replace the last line of the code fragment with `sns.pairplot(dfDataAnt13, diag_kind="kde")`, we get a new visualization, with a better view of the distribution. This is shown in *Figure 6.3*:

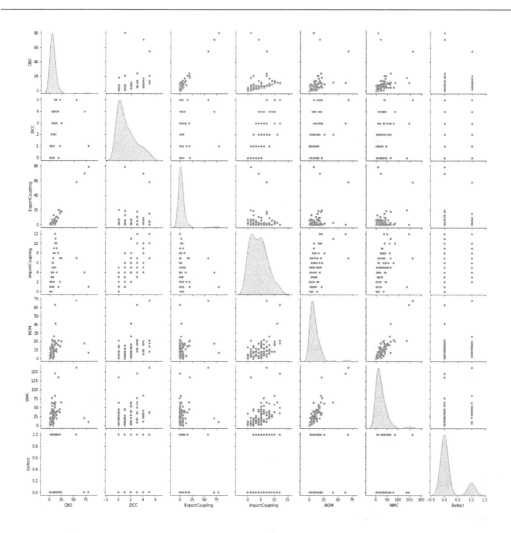

Figure 6.3 – Correlogram with a better visualization of the distribution of each measure

These correlograms provide us with a quick orientation of which variables can be correlated with one another. These correlations are something that we can use later in our work.

We can also look at the data by visualizing the numbers using heatmaps. Heatmaps are tabular visualizations where the intensity of the color indicates the strength of the value of each variable. We can use the following code to create a heatmap:

```
# heatmap
p1 = sns.heatmap(dfDataAnt13, cmap="Reds")
```

The resulting diagram is presented in *Figure 6.4*:

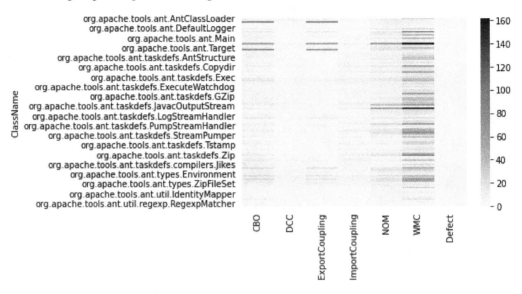

Figure 6.4 – Summary heatmap for the measures

Before diving deeper into correlation analysis, I often dive a bit deeper into pairwise comparisons. I also recommend my students to do that since working with pairs allows us to understand the connections between variables.

So, here is my next best practice.

> **Best practice #35**
> When visualizing data at the aggregate level, focus on the strength of relationships and connections between the values.

Visualizations at the aggregate level can provide us with many different views, but what we should look for is the connections between the variables. Correlograms and heatmaps provide us with this kind of visualization and understanding of the data.

Diving deeper into correlations

A good set of diagrams to work with are scatter plots. However, I often use diagrams that are called KDE plots, also called *density plots*. They provide a nicer overview of the variables. The following code fragment visualizes the data in this way:

```
# now, let's make some density plots
# set seaborn style
```

```
sns.set_style("white")

# Basic 2D density plot
sns.kdeplot(x=dfDataAnt13.CBO, y=dfDataAnt13.DCC)
plt.show()
```

The result of this code fragment is the diagram presented in *Figure 6.5*:

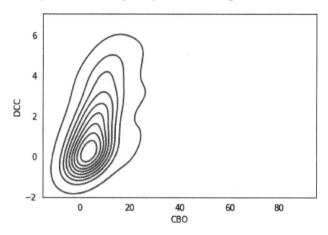

Figure 6.5 – Density plot for two measures – DCC and CBO

This diagram indicates that two measures – CBO and DCC – are quite strongly dependent on one another (or they quantify similar/same measurable concepts).

We can make this diagram a bit nicer if we want to use it in a dashboard by using the following code fragment:

```
# Custom the color, add shade and bandwidth
sns.kdeplot(x=dfDataAnt13.WMC,
            y=dfDataAnt13.ImportCoupling,
            cmap="Reds",
            shade=True,
            bw_adjust=.5)
plt.show()
```

This code fragment results in the following diagram:

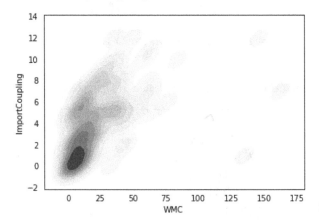

Figure 6.6 – Density plot with a colormap

The preceding diagram shows both the correlation and the number of points that are in each area – the more intense the color, the more data points are in that area. The same diagram for the DCC and CBO measures is shown in *Figure 6.7*:

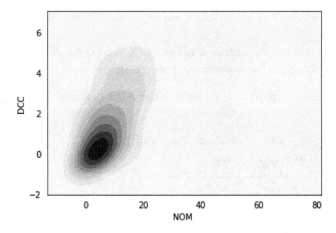

Figure 6.7 – Density plot with a colormap for the DCC and CBO measures

Finally, we can use bubble diagrams to visualize the correlations and the number of data points per group. The following code creates the bubble diagram:

```
# now a bubble diagram
# use the scatterplot function to build the bubble map
sns.scatterplot(data=dfDataAnt13,
```

```
          x="NOM",
          y="DCC",
          size="Defect",
          legend=False,
          sizes=(20, 2000))

# show the graph
plt.show()
```

This code results in the diagram presented in *Figure 6.8*:

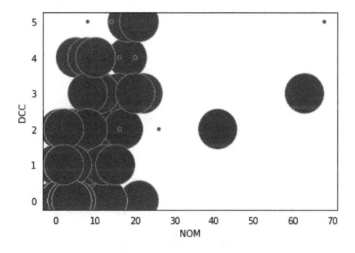

Figure 6.8 – Scatter plot – a variant called a bubble plot

This plot lets us see the number of points in each area of the scatterplot, which helps us track the correlations visually.

Summarizing individual measures

Scatterplots and density plots are good for tracking dependencies between variables. However, we often need to summarize individual measures. For that, we can use boxplots. The following code creates a boxplot for the data in our example:

```
# boxplot
sns.boxplot( x=dfDataAnt13.Defect, y=dfDataAnt13.CBO )
```

The result is the boxplot presented in *Figure 6.9*:

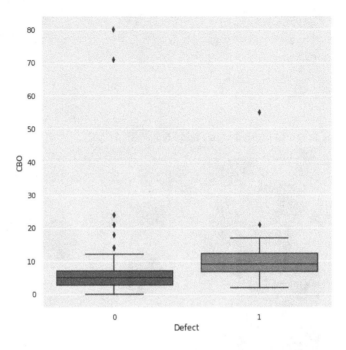

Figure 6.9 – Boxplot summarizing the CBO measure for classes with and without defects

The summary provides us with a quick visual indication that the classes with defects are often more connected to other classes than the classes without defects. This is not that surprising because usually, the classes are connected and the ones that are *not* connected are often very simple and therefore not error-prone.

A variation of the boxplot is the violin plot, which we get if we change the last line of the last code fragment to `sns.violinplot( x='Defect', y='CBO', data=dfDataAnt13)`. *Figure 6.10* presents such a violin diagram:

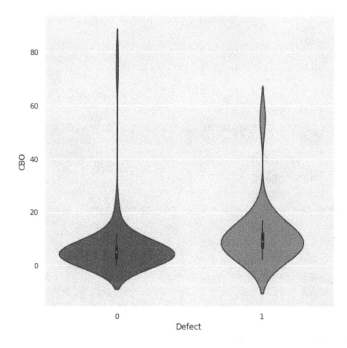

Figure 6.10 – A violin diagram, which is a variation of a boxplot

Visualization is a good way to understand the numerical data that we have at our disposal. We can go even further and start working with it by using methods such as dimensionality reduction.

So, here is the next best practice.

Best practice #36
Diving deeper into individual analyses should be guided by the machine learning task at hand.

Although we have not discussed the tasks for our numerical data explicitly, it is always there. In the case of defect-related data, the most common task is predicting the number of defects per module or class. This means that charts such as the violin plot are very useful, providing us with a visual understanding of whether there is some sort of difference – a difference that a machine learning model can capture.

Reducing the number of measures – PCA

The final analysis that we'll do regarding numerical data in this chapter is about reducing the number of variables. It comes from the field of statistics and has been used for reducing the number of variables in the experiment: PCA (Wold, 1987 #104). In short, PCA is a technique that finds the best possible fit of a pre-defined number of vectors to the data at hand. It does not remove any variables; instead, it recalculates them in such a way that the correlation among the new set of variables – called principal components – is minimalized.

Let's apply this to our dataset using the following code fragment:

```
# before we use PCA, we need to remove the variable "defect"
# as this is the variable which we predict
dfAnt13NoDefects = dfDataAnt13.drop(['Defect'], axis=1)


# PCA for the data at hand
from sklearn.decomposition import PCA


# we instantiate the PCA class with two parameters
# the first one is the number of principal components
# and the second is the random state
pcaComp = PCA(n_components=2,
              random_state=42)


# then we find the best fit for the principal components
# and fit them to the data
vis_dims = pcaComp.fit_transform(dfAnt13NoDefects)
```

Now, we can visualize the data:

```
# and of course, we could visualize it
import matplotlib.pyplot as plt
import matplotlib
import numpy as np


colors = ["red", "darkgreen"]
x = [x for x,y in vis_dims]
y = [y for x,y in vis_dims]


# please note that we use the dataset with defects to
# assign colors to the data points in the diagram
color_indices = dfDataAnt13.Defect


colormap = matplotlib.colors.ListedColormap(colors)
plt.scatter(x, y, c=color_indices, cmap=colormap, alpha=0.3)
for score in [0,1]:
    color = colors[score]
plt.rcParams['figure.figsize'] = (20,20)
```

This code fragment results in the diagram presented in *Figure 6.11*:

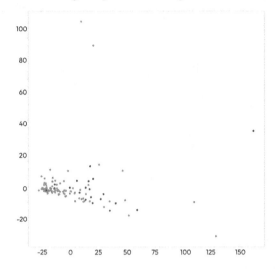

Figure 6.11 – PCA results for reducing the dimensionality of the defect dataset to two. The red data points are the classes that have defects and the green data points are the classes without defects

What is typical about a PCA transformation is its linearity. We can see that this diagram contains trails of that – it looks like a triangle with one horizontal dimension along the *x axis*, one vertical dimension along the *y axis*, and a 0-point on the left-hand side.

For this dataset, the diagram shows that the red-marked data points are grouped toward the left, and the green-marked points are spread a bit more to the right. This means that there is some difference between the classes that have defects and the classes that do not have defects. However, there is no clear-cut distinction. That would indicate that the machine learning model can't find a pattern – at least, not a pattern that would be robust.

Other types of data – images

In *Chapter 3*, we looked at image data, mostly from the perspective of what kind of image data exists. Now, we will take a more pragmatic approach and introduce a better way of working with images than just using files.

Let's look at how image data is stored in a popular repository – Hugging Face. The library has a specific module for working with datasets – conveniently called *Dataset*. It can be installed using the `pip install -q datasets` command. So, let's load a dataset and visualize one of the images from there using the following code fragment:

```
# importing the images library
from datasets import load_dataset, Image
```

```
# loading a dataset "food101", or more concretely it's split for
training
dataset = load_dataset("food101", split="train")
```

Now, the variable dataset contains all the images. Well, not all of them – just the part that the designer of the dataset specified as the training set (see the last line of the code fragment). We can visualize one of the images using the following code:

```
# visualizing the first image
dataset[0]["image"]
```

Since the images are under unknown copyright, we won't visualize them in this book. However, the preceding line will show the first image in the dataset. We can also take a look at what else is in that dataset by simply typing dataset. We will see the following output:

```
Dataset({ features: ['image', 'label'], num_rows: 75750 })
```

This means that the dataset contains two columns – images and their labels. It contains 75,750 of them. Let's see what the distribution of labels looks like in this dataset using the following code:

```
# we can also plot the histogram
# to check the distribution of labels in the dataset
import seaborn as sns
import matplotlib.pyplot as plt

plt.rcParams['figure.figsize'] = (20,10)
sns.histplot(data=dataset['label'], x=dataset['label'])
```

This gives us a nice histogram, as shown in *Figure 6.12*:

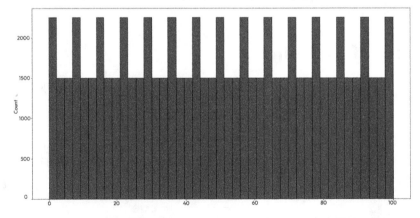

Figure 6.12 – Histogram with the distributions of labels. Each column is the
number of images labeled with the appropriate label – 0 to 100

This diagram shows classes of images that are larger than others – the ones that contain over 2,000 images in them. However, it is difficult to check what these labels mean without understanding the dataset. We can do that by manually visualizing the images. So, here is my next best practice.

> **Best practice #37**
> When visualizing the metadata for images, make sure you visualize the images themselves.

We mustn't forget to visualize the image data by plotting the images. We need to ensure that we know what the labels mean and what we use them for.

Text data

For the text data, we'll use the same Hugging Face hub to obtain two kinds of data – unstructured text, as we did in *Chapter 3*, and structured data – programming language code:

```
# import Hugging Face Dataset
from datasets import load_dataset

# load the dataset with text classification labels
dataset = load_dataset('imdb')
```

The preceding code fragment loads the dataset of movie reviews from the **Internet Movie Database** (**IMDb**). We can get an example of the data by using an interface that's similar to what we used for images:

```
# show the first example
dataset['train'][0]
```

We can visualize it using a similar one too:

```
# plot the distribution of the labels
sns.histplot(dataset['train']['label'], bins=2)
```

The preceding code fragment creates the following diagram, showing that both positive and negative comments are perfectly balanced:

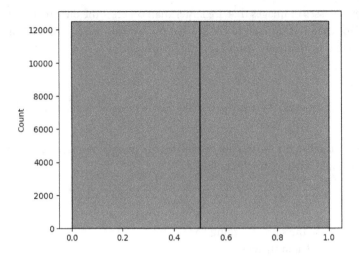

Figure 6.13 – Balanced classes in the IMDb movie database reviews

We can do all kinds of processing on the text data in the next steps. However, this processing is related to feature extraction, so we'll talk about it in the next few chapters.

Before we do that, though, let's look at datasets that are closer to the domain of software engineering – programming language code. We used similar data in *Chapter 3*, so let's focus on how we could obtain a larger corpus of programming language code from Hugging Face. We could use the following code to obtain the data and check the first program:

```
# now, let us import the code to the text summarization dataset
dsCode = load_dataset('code_x_glue_ct_code_to_text', 'java',
split='test')


# and see the first example of the code
dsCode[0]
```

This code fragment shows us the first program, which is already tokenized and prepared for further analysis. So, let's take a peek at the frequency of tokens in this dataset. We can use the following code for that:

```
import pandas as pd
import matplotlib.pyplot as plt


# create a list of tokens
lstCodeLines = dsCode['code_tokens']


# flatten the list of lists to one list
lstCodeLines = [item for sublist in lstCodeLines for item in sublist]
```

```
#print the first elements of the list
print(lstCodeLines[:10])

dfCode = pd.DataFrame(lstCodeLines, columns=['token'])

# group the tokens and count the number of occurences
# which will help us to visualize the frequency of tokens in the next
step
dfCodeCounts = dfCode.groupby('token').size().reset_
index(name='counts')

# sort the counts by descending order
dfCodeCounts = dfCodeCounts.sort_values(by='counts', ascending=False)

fig, ax = plt.subplots(figsize=(12, 6))

# plot the frequency of tokens as a barplot
# for the simplicity, we only take the first 20 tokens
sns.barplot(x='token',
            y='counts',
            data=dfCodeCounts[:20],
            palette=sns.color_palette("BuGn_r", n_colors=20),
            ax=ax)

# rotate the x-axis labels to make sure that
# we see the full token names, i.e. lines of code
ax.set_xticklabels(ax.get_xticklabels(),
                   rotation=45,
                   horizontalalignment='right')
```

The preceding code extracts the tokens, counts them, and creates a diagram of the frequency of the top 20 tokens. The result is presented in *Figure 6.14*:

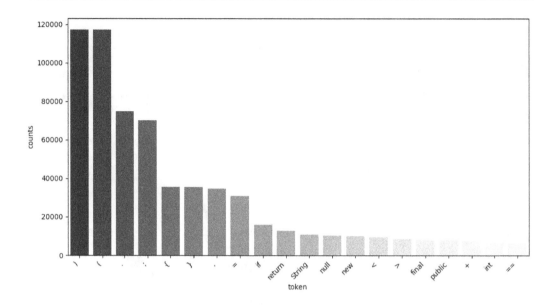

Figure 6.14 – Token frequencies for the top 20 most common tokens in the code dataset

Interestingly, we can observe that brackets, commas, semicolons, and curly brackets are the most commonly used tokens in the dataset. This isn't surprising as these kinds of characters have special meaning in Java.

The other tokens in the top 20 list are, unsurprisingly, keywords in Java or have special meanings (such as ==).

So, my last best practice in this chapter is about understanding the text data.

> **Best practice #38**
> Summary statistics for text data help us perform a sanity check of the data.

Even though textual data is quite unstructured by nature, we can visualize some of the properties of the data. For example, token frequency analysis can reveal whether our empirical understanding of the data makes sense and whether we can trust it.

Toward feature engineering

In this chapter, we explored methods for visualizing data. We learned how to create diagrams and identify dependencies in the data. We also learned how we can use dimensionality reduction techniques to plot multidimensional data on a two dimensional diagram.

In the next few chapters, we'll dive into feature engineering different types of data. Sometimes, it is easy to mix feature engineering with data extraction. In practice, it is not that difficult to tell one from the other.

Extracted data is data that has been collected by applying some sort of measurement instrument. Raw text or images are good examples of this kind of data. Extracted data is close to the domain where the data comes from – or how it is measured.

Features describe the data based on the analysis that we want to perform – they are closer to what we want to do with the data. It is closer to what we want to achieve and which form of machine learning analysis we want to do.

References

- *International Standardization Organization, International vocabulary of basic and general terms in metrology (VIM). In International Organization. 2004. p. 09-14.*

- *Alhusain, S. Predicting Relative Thresholds for Object Oriented Metrics. In 2021 IEEE/ACM International Conference on Technical Debt (TechDebt). 2021. IEEE.*

- *Feldt, R., et al. Supporting software decision meetings: Heatmaps for visualising test and code measurements. In 2013 39th Euromicro Conference on Software Engineering and Advanced Applications. 2013. IEEE.*

- *Staron, M., et al. Measuring and visualizing code stability – a case study at three companies. In 2013 Joint Conference of the 23rd International Workshop on Software Measurement and the 8th International Conference on Software Process and Product Measurement. 2013. IEEE.*

- *Wen, S., C. Nilsson, and M. Staron. Assessing the release readiness of engine control software. In Proceedings of the 1st International Workshop on Software Qualities and Their Dependencies. 2018.*

7

Feature Engineering for Numerical and Image Data

In most cases, when we design large-scale machine learning systems, the types of data we get require more processing than just visualization. This visualization is only for the design and development of machine learning systems. During deployment, we can monitor the data, as we discussed in the previous chapters, but we need to make sure that we use optimized data for inference.

Therefore, in this chapter, we'll focus on feature engineering – finding the right features that describe our data closer to the problem domain rather than closer to the data itself. Feature engineering is a process where we extract and transform variables from raw data so that we can use them for predictions, classifications, and other machine learning tasks. The goal of feature engineering is to analyze and prepare the data for different machine learning tasks, such as making predictions or classifications.

In this chapter, we'll focus on the feature engineering process for numerical and image data. We'll start by going through the typical methods, such as **principal component analysis** (**PCA**), which we used previously for visualization. Then, we'll cover more advanced methods, such as **t-student distribution stochastic network embedding** (**t-SNE**) and **independent component analysis** (**ICA**). What we'll end up with is the use of autoencoders as a dimensionality reduction technique for both numerical and image data.

In this chapter, we'll cover the following topics:

- Feature engineering process fundamentals
- PCA and similar methods
- Autoencoders for numerical and image data

Feature engineering

Feature engineering is the process of transforming raw data into numerical values that can be used in machine learning algorithms. For example, we can transform raw data about software defects (for example, their description, the characteristics of the module they come from, and so on) into a table of numerical values that we can use for machine learning. The raw numerical values, as we saw in the previous chapter, are the result of quantifying entities that we use as sources of data. They are the results of applying measurement instruments to the data. Therefore, by definition, they are closer to the problem domain rather than the solution domain.

The features, on the other hand, quantify the raw data and contain only the information that is important for the machine learning task at hand. We use these features to make sure that we find the patterns in the data during training that we can then use during deployment. If we look at this process from the perspective of measurement theory, this process changes the abstraction level of the data. If we look at this process from a statistical perspective, this is the process of removing noise and reducing the dimensions of the data.

In this chapter, we'll focus on the process of reducing the dimensions of the data and denoising the image data using advanced methods such as autoencoders.

Figure 7.1 presents where feature extraction is placed in a typical machine learning pipeline. This pipeline was presented in *Chapter 2*:

Figure 7.1 – Feature engineering in a typical machine learning pipeline

This figure shows that the features are as close to the clean and validated data as possible, so we need to rely on the techniques from the previous chapters to visualize the data and reduce the noise. The next activity, after feature engineering, is modeling the data, as presented in *Figure 7.2*. This figure shows a somewhat simplified view of the entire pipeline. This was also presented in *Chapter 2*:

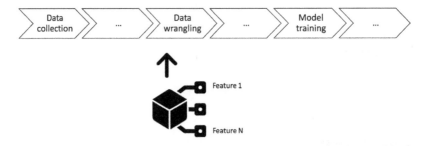

Figure 7.2 – A typical machine learning pipeline. A somewhat simplified view from Chapter 2

We covered modeling previously, so let's dive deeper into the feature extraction process. Since numerical and image data are somewhat similar from this perspective, we'll discuss them together in this chapter. Text data is different and therefore we have devoted the next chapter to it.

My first best practice in this chapter, however, is related to the link between feature extraction and models.

> **Best practice #39**
>
> Use feature engineering techniques if the data is complex, but the task is simple – for example, creating a classification model.

If the data is complex and the task is complex, try to use complex but capable models, such as the transformer models presented later in this book. An example of such a task can be code completion when the model finished creating a piece of a program that a programmer started to write. Simplifying complex data for simpler models allows us to increase the explainability of the trained models because we, as AI engineers, are more involved in the process through data wrangling.

Feature engineering for numerical data

We'll introduce feature engineering for numerical data by using the same technique that we used previously but for visualizing data – PCA.

PCA

PCA is used to transform a set of variables into components that are supposed to be independent of one another. The first component should explain the variability of the data or be correlated with most of the variables. *Figure 7.3* illustrates such a transformation:

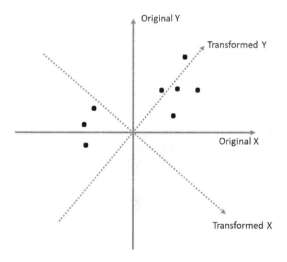

Figure 7.3 – Graphical illustration of the PCA transformation from two dimensions to two dimensions

This figure contains two axes – the blue ones, which are the original coordinates, and the orange ones, which are the imaginary axes and provide the coordinates for the principal components. The transformation does not change the values of the *x* and *y* axes and instead finds such a transformation that the axes align with the data points. Here, we can see that the transformed *Y* axis aligns better with the data points than the original *Y* axis.

Now, let's execute a bit of code that can read the data and make such a PCA transformation. In this example, the data has six dimensions – that is, six variables:

```
# read the file with data using openpyxl
import pandas as pd

# we read the data from the excel file,
# which is the defect data from the ant 1.3 system
dfDataAnt13 = pd.read_excel('./chapter_6_dataset_numerical.
            xlsx',sheet_name='ant_1_3', index_col=0)
dfDataAnt13
```

The preceding code fragment reads the data and shows that it has six dimensions. Now, let's create the PCA transformation. First, we must remove the dependent variable in our dataset – `Defect`:

```
# let's remove the defect column, as this is the one that
# we could potentially predict
dfDataAnt13Pred = dfDataAnt13.drop(['Defect'], axis = 1)
```

Then, we must import the PCA transformation and execute it. We want to find a transformation from the five variables (six minus the `Defect` variable) to three dimensions. The number of dimensions is completely arbitrary, but since we used two in the previous chapters, let's use more this time:

```
# now, let's import PCA and find a few components
from sklearn.decomposition import PCA

# previously, we used 2 components, now, let's go with
# three
pca = PCA(n_components=3)

# now, the transformation to the new components
dfDataAnt13PCA = pca.fit_transform(dfDataAnt13Pred)

# and printing the resulting array
# or at least the three first elements
dfDataAnt13PCA[:3]
```

The resulting DataFrame – `dfDataAnt13PCA` – contains the values of the transformed variables. They are as independent from one another as possible (linearly independent).

I would like to emphasize the general scheme of how we work with this kind of data transformation because that is a relatively standard way of doing things.

First, we instantiate the transformation module and provide the arguments. In most cases, the arguments are plenty, but there is one, n_components, that describes how many components we want to have.

Second, we use the fit_transform() function to train the classifier and transform it into these components. We use these two operations together, simply because these transformations are data-specific. There is no need to train the transformation on one data and apply it to another one.

What we can also do with PCA, which we cannot do with other types of transformations, is check how much variability each component explains – that is, how well the components align with the data. We can do this with the following code:

```
# and let's visualize that using the seaborn library
import seaborn as sns

sns.set(rc={"figure.figsize":(8, 8)})
sns.set_style("white")
sns.set_palette('rocket')

sns.barplot(x=['PC 1', 'PC 2', 'PC 3'], y=pca.explained_variance_
ratio_)
```

This code fragment results in the diagram presented in *Figure 7.4*:

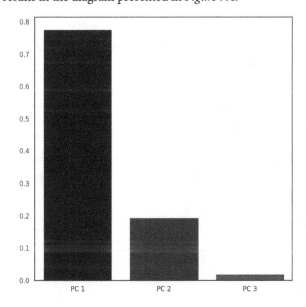

Figure 7.4 – Variability explained by the principal components

This figure shows that the first component is the most important one – that is, it explains the largest amount of variability. This variability can be seen as the amount of information that the data contains. In the case of this dataset, the first component explains about 80% of the variability and the second one almost 20%. This means that our dataset has one dominating dimension and some dispersion of the data along a second dimension. The third dimension is almost non-existent.

This is where my next best practice comes in.

> **Best practice #40**
> Use PCA if the data is somehow linearly separable and on similar scales.

If the data is linear, or multilinear, PCA makes a large difference for training the model. However, if the data is not linear, use a more complex model, such as t-SNE.

t-SNE

As a transformation, PCA works well when data is linearly separable to some extent. In practice, this means that the coordinate system can be positioned in such a way that most of the data is on one of its axes. However, not all data is like that. One example of data that is not like that is data that can be visualized as a circle – it is equally distributed along both axes.

To reduce the dimensions of non-linear data, we can use another technique – t-SNE. This kind of dimensionality reduction technique is based on extracting the activation values of a neural network, which is trained to fit the input data.

The following code fragment creates such a t-SNE transformation of the data. It follows the same schema that was described for the PCA and it also reduces the dimensions to three:

```
# for t-SNE, we use the same data as we used previously
# i.e., the predictor dfDataAnt13Pred
from sklearn.manifold import TSNE

# we create the t-sne transformation with three components
# just like we did with the PCA
tsne = TSNE(n_components = 3)

# we fit and transform the data
dfDataAnt13TSNE = tsne.fit_transform(dfDataAnt13Pred)

# and print the three first rows
dfDataAnt13TSNE[:3]
```

The resulting DataFrame – `dfDataAnt13TSNE` – contains the transformed data. Unfortunately, the t-SNE transformation does not allow us to get the value of the explained variability, simply because this concept does not exist for such a transformation. However, we can visualize it. The following figure presents a 3D projection of the three components:

Ant 1.3 modules and their defect proneness

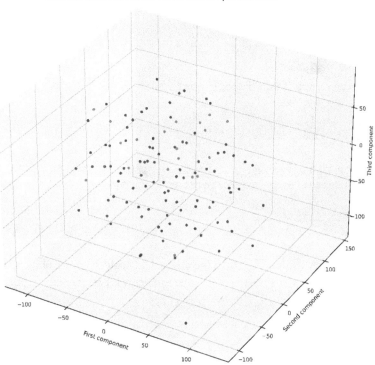

Figure 7.5 – Visualization of the t-SNE components. Green dots represent defect-free components and red dots represent components with defects

Here is my next best practice for this chapter.

> **Best practice #41**
> Use t-SNE if you do not know the properties of the data and the dataset is large (more than 1,000 data points).

t-SNE is a very good and robust transformation. It works particularly well for large datasets – that is, those that consist of hundreds of data points. One of the challenges, however, is that there is no interpretation of the components that t-SNE delivers. We should also know that the best results from t-SNE require hyperparameters to be tuned carefully.

ICA

We can use another kind of transformation here – ICA. This transformation works in such a way that it finds the least correlated data points and separates them. It's been historically used in the medical domain to remove disturbances and artifacts from high-frequency **electroencephalography** (**EEG**) signals. An example of such a disturbance is the 50 - Hz electrical power signal.

However, it can be used for any kind of data. The following code fragment illustrates how ICA can be used for the same dataset that we used in the previous transformations:

```
# we import the package
from sklearn.decomposition import FastICA

# instantiate the ICA
ica = FastICA(n_components=3)

# transform the data
dfDataAnt13ICA = ica.fit_transform(dfDataAnt13Pred)

# and check the first three rows
dfDataAnt13ICA[:3]
```

ICA needs to result in fewer components than the original data, although we only used three in the preceding code fragment. The visualization of these components is presented in the following figure:

Ant 1.3 modules and their defect proneness

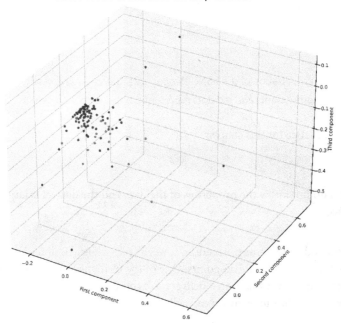

Figure 7.6 – Visualization of the dataset transformed using ICA

In *Figure 7.6*, green components are the ones without defects and the red ones contain defects.

Locally linear embedding

A technique that is somewhat in between t-SNE and PCA (or ICA) is known as **locally linear embedding** (**LLE**). This technique assumes that neighboring nodes are placed close to one another on some kind of virtual plane. The algorithm trains a neural network in such a way that it preserves the distances between the neighboring nodes.

The following code fragment illustrates how to use the LLE technique:

```
from sklearn.manifold import LocallyLinearEmbedding
# instantiate the classifier
lle = LocallyLinearEmbedding(n_components=3)

# transform the data
dfDataAnt13LLE = lle.fit_transform(dfDataAnt13Pred)

# print the three first rows
dfDataAnt13LLE[:3]
```

This fragment results in a similar DataFrame to the previous algorithms. Here is the visualization:

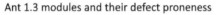

Ant 1.3 modules and their defect proneness

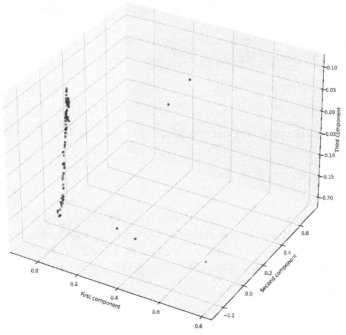

Figure 7.7 – Visualization of the LLE components

All the techniques we've discussed so far are flexible and allow us to indicate how many components we need in the transformed data. However, sometimes, the problem is that we do not know how many components we need.

Linear discriminant analysis

Linear discriminant analysis (LDA) is a technique that results in as many components as we have in our dataset. This means that the number of columns in our dataset is the same as the number of components that LDA provides. This, in turn, means that we need to define one of the variables as the dependent one for the algorithm to work.

The LDA algorithm finds a projection of the dataset on a lower dimensional space in such a way that it separates the data in the classes of the dependent variable. Therefore, we need one. The following code fragment illustrates the use of LDA on our dataset:

```
from sklearn.discriminant_analysis import LinearDiscriminantAnalysis

# create the classifier
# please note that we can only use one component, because
# we have only one predicted variable
lda = LinearDiscriminantAnalysis(n_components=1)

# fit to the data
# please note that this transformation requires the predicted
# variable too
dfDataAnt13LDA = lda.fit(dfDataAnt13Pred, dfDataAnt13.Defect).
transform(dfDataAnt13Pred)

# print the transformed data
dfDataAnt13LDA[:3]
```

The resulting DataFrame contains only one component as we have only one dependent variable in our dataset.

Autoencoders

In recent years, a new technique for feature extraction has been gaining popularity – autoencoders. Autoencoders are special kinds of neural networks that are designed to transform data from one type of data into another. Usually, they are used to recreate the input data in a slightly modified form. For example, they can be used to remove noise from images or change images to use different styles of brushes.

Autoencoders are quite generic and can be used for other kinds of data, which we'll learn about in the remainder of this chapter (for example, for image data). *Figure 7.8* presents the conceptual model of autoencoders. They consist of two parts – an encoder and a decoder. The role of the encoder is to transform the input data – an image in this example – into an abstract representation. This abstract representation is stored in a specific layer (or layers), which is called the bottleneck. The role of the bottleneck is to store such properties of the input data that allow the decoder to recreate the data. The role of the decoder is to take the abstract representation of the data from the bottleneck layer and re-create the input data as best as possible:

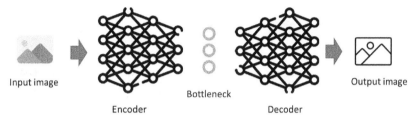

Input image   Bottleneck   Output image

Encoder   Decoder

Figure 7.8 – Conceptual visualization of an autoencoder. Here, the input data is in the form of an image

Since autoencoders are trained to recreate the data as best as possible, the bottleneck values are generally believed to be a good internal representation of the input data. It is such a good representation that it allows us to discriminate between different input data points.

The bottleneck values are also very flexible. As opposed to the techniques presented previously, there is no limit on how many features we can extract. If we need to, we can even extract more features than we have columns in our dataset, although it does not make much sense.

So, let's construct a pipeline for extracting features from an autoencoder that's designed to learn the representation of the defect data.

The following code fragment illustrates reading the dataset and removing the defect column from it:

```python
# read the file with data using openpyxl
import pandas as pd


# we read the data from the excel file,
# which is the defect data from the ant 1.3 system
dfDataAnt13 = pd.read_excel('./chapter_6_dataset_numerical.
            xlsx',sheet_name='ant_1_3',index_col=0)
# let's remove the defect column, as this is the one that we could
# potentially predict
X = dfDataAnt13.drop(['Defect'], axis = 1)
y = dfDataAnt13.Defect
```

In addition to the removal of the column, we need to scale the data so that the autoencoder has a good chance of recognizing small patterns in all columns:

```
# split into train test sets
X_train, X_test, y_train, y_test = train_test_split(X, y, test_
size=0.33, random_state=1)

# scale data
t = MinMaxScaler()
t.fit(X_train)
X_train = t.transform(X_train)
X_test = t.transform(X_test)
```

Now, we can create the encoder part of our autoencoder, which is shown in the following code fragment:

```
# number of input columns
n_inputs = X.shape[1]

# the first layer - the visible one
visible = Input(shape=(n_inputs,))

# encoder level 1
e = Dense(n_inputs*2)(visible)
e = BatchNormalization()(e)
e = LeakyReLU()(e)

# encoder level 2
e = Dense(n_inputs)(e)
e = BatchNormalization()(e)
e = LeakyReLU()(e)
```

The preceding code creates two levels of the autoencoder since our data is quite simple. Now, the interesting part is the bottleneck, which can be created by running the following code:

```
n_bottleneck = 3
bottleneck = Dense(n_bottleneck)(e)
```

In our case, the bottleneck is very narrow – only three neurons – as the dataset is rather small and it is not very complex. In the next part, when we use the autoencoder for images, we will see that the bottleneck can be much wider. The general idea is that wider bottlenecks allow us to capture more complex dependencies in the data. For example, for color images, we need more neurons as we need to capture colors, while for grayscale images, we need narrower bottlenecks.

Finally, we can create the decoder part of the autoencoder by using the following code:

```
# define decoder, level 1
d = Dense(n_inputs)(bottleneck)
```

```
d = BatchNormalization()(d)
d = LeakyReLU()(d)
# decoder level 2
d = Dense(n_inputs*2)(d)
d = BatchNormalization()(d)
d = LeakyReLU()(d)
# output layer
output = Dense(n_inputs, activation='linear')(d)
```

The last part of the construction process is to put these three parts together – the encoder, the bottleneck, and the decoder. We can use the following code for this:

```
# we place both of these into one model
# define autoencoder model
model = Model(inputs=visible, outputs=output)
# compile autoencoder model
model.compile(optimizer='adam', loss='mse')
```

At this point, we have constructed our autoencoder. We've defined its layers and the bottleneck. Now, the autoencoder must be trained to understand how to represent our data. We can do this using the following code:

```
# we train the autoencoder model
history = model.fit(X_train, X_train,
                    epochs=100,
                    batch_size=16,
                    verbose=2,
                    validation_data=(X_test,X_test))
```

Please note that we use the same data as input and as validation since we need to train the encoder to re-create the same data as accurately as possible, given the size of the bottleneck. After training the encoder model, we can use it to extract the bottleneck values from the model. We can do this by defining a submodel and using it for input data:

```
submodel = Model(model.inputs, model.get_layer("dense_8").output)

# this is the actual feature extraction -
# where we make prediction for the train dataset
# please note that the autoencoder requires a two dimensional array
# so we need to take one datapoint and make it into a two dimensional
array
# with only one row
results = submodel.predict(np.array([X_train[0]]))

results[0]
```

The outcome of executing this code is a vector of three values – the bottleneck values of the autoencoder.

My next best practice in this chapter is related to the use of autoencoders.

> **Best practice #42**
>
> Use autoencoders for numerical data when the dataset is really large since autoencoders are complex and require a lot of data for training.

Since the quality of the features is a function of how well the autoencoder is trained, we need to make sure that the training dataset is large. Therefore, autoencoders are often used for image data.

Feature engineering for image data

One of the most prominent feature extraction methods for image data is the use of **convolutional neural networks** (**CNNs**) and extracting embeddings from these networks. In recent years, a new type of this kind of neural network was introduced – autoencoders. Although we can use autoencoders for all kinds of data, they are particularly well-suited for images. So, let's construct an autoencoder for the MNIST dataset and extract bottleneck values from it.

First, we need to download the MNIST dataset using the following code fragment:

```
# first, let's read the image data from the Keras library
from tensorflow.keras.datasets import mnist

# and load it with the pre-defined train/test splits
(X_train, y_train), (X_test, y_test) = mnist.load_data()
X_train = X_train/255.0
X_test = X_test/255.0
```

Now, we can construct the encoder part by using the following code. Please note that there is one extra layer in the encoder part. The goal of this layer is to transform a two-dimensional image into a one-dimensional input array – to flatten the image:

```
# image size is 28 pixels
n_inputs = 28

# the first layer - the visible one
visible = Input(shape=(n_inputs,n_inputs,))

# encoder level 1
e = Flatten(input_shape = (28, 28))(visible)
e = LeakyReLU()(e)
e = Dense(n_inputs*2)(e)
e = BatchNormalization()(e)
```

```
e = LeakyReLU()(e)

# encoder level 2
e = Dense(n_inputs)(e)
e = BatchNormalization()(e)
e = LeakyReLU()(e)
```

Now, we can construct our bottleneck. In this case, the bottleneck can be much wider as the images are more complex (and there are more of them) than the array of numerical values of modules, which we used in the previous autoencoder:

```
n_bottleneck = 32
bottleneck = Dense(n_bottleneck)(e)
```

The decoder part is very similar to the previous example, with one extra layer that re-creates the image from its flat representation:

```
# and now, we define the decoder part
# define decoder, level 1
d = Dense(n_inputs)(bottleneck)
d = BatchNormalization()(d)
d = LeakyReLU()(d)
# decoder level 2
d = Dense(n_inputs*2)(d)
d = BatchNormalization()(d)
d = LeakyReLU()(d)
# output layer
d = Dense(n_inputs*n_inputs, activation='linear')(d)
output = Reshape((28,28))(d)
```

Now, we can compile and train the autoencoder:

```
# we place both of these into one model
# define autoencoder model
model = Model(inputs=visible, outputs=output)
# compile autoencoder model
model.compile(optimizer='adam', loss='mse')

# we train the autoencoder model
history = model.fit(X_train, X_train,
                    epochs=100,
                    batch_size=16,
                    verbose=2,
                    validation_data=(X_test,X_test))
```

Finally, we can extract the bottleneck values from the model:

```
submodel = Model(model.inputs, bottleneck)

# this is the actual feature extraction -
# where we make prediction for the train dataset
# please note that the autoencoder requires a two dimensional array
# so we need to take one datapoint and make it into a two dimensional
array
# with only one row
results = submodel.predict(np.array([X_train[0]]))

results[0]
```

Now, the resulting array of values is much larger – it has 32 values, the same number of neurons that we have in our bottleneck.

The number of neurons in the bottleneck is essentially arbitrary. Here's a best practice for selecting the number of neurons.

> **Best practice #43**
>
> Start with a small number of neurons in the bottleneck – usually one third of the number of columns. If the autoencoder does not learn, increase the number gradually.

There is no specific reason why I chose 1/3rd of the number of columns, just experience. You can start from the opposite direction – make the bottleneck layer as wide as the input – and decrease gradually. However, having the same number of features as the number of columns is not why we use feature extraction in the first place.

Summary

In this chapter, our focus was on feature extraction techniques. We explored how we can use dimensionality reduction techniques and autoencoders to reduce the number of features in order to make machine learning models more effective.

However, numerical and image data are only two examples of data. In the next chapter, we continue with the feature engineering methods, but for textual data, which is more common in contemporary software engineering.

References

- Zheng, A. and A. Casari, *Feature engineering for machine learning: principles and techniques for data scientists.* 2018: O'Reilly Media, Inc

- Heaton, J. *An empirical analysis of feature engineering for predictive modeling. In SoutheastCon 2016.* 2016. IEEE.

- Staron, M. and W. Meding, *Software Development Measurement Programs. Springer. https://doi.org/10.1007/978-3-319-91836-5. Vol. 10.* 2018. 3281333.

- Abran, A., *Software metrics and software metrology.* 2010: John Wiley & Sons.

- Meng, Q., et al. *Relational autoencoder for feature extraction. In 2017 International joint conference on neural networks (IJCNN).* 2017. IEEE.

- Masci, J., et al. *Stacked convolutional auto-encoders for hierarchical feature extraction. In Artificial Neural Networks and Machine Learning, ICANN 2011: 21st International Conference on Artificial Neural Networks, Espoo, Finland, June 14-17, 2011, Proceedings, Part I 21.* 2011. Springer.

- Rumelhart, D.E., G.E. Hinton, and R.J. Williams, *Learning representations by back-propagating errors. nature,* 1986. 323(6088): p. 533-536.

- Mosin, V., et al., *Comparing autoencoder-based approaches for anomaly detection in highway driving scenario images. SN Applied Sciences,* 2022. 4(12): p. 334.

8

Feature Engineering for Natural Language Data

In the previous chapter, we explored how to extract features from numerical data and images. We explored a few algorithms that are used for that purpose. In this chapter, we'll continue with the algorithms that extract features from natural language data.

Natural language is a special kind of data source in software engineering. With the introduction of GitHub Copilot and ChatGPT, it became evident that machine learning and artificial intelligence tools for software engineering tasks are no longer science fiction. Therefore, in this chapter, we'll explore the first steps that made these technologies so powerful – feature extraction from natural language data.

In this chapter, we'll cover the following topics:

- Tokenizers and their role in feature extraction
- Bag-of-words as a simple technique for processing natural language data
- Word embeddings as more advanced methods that can capture contexts

Natural language data in software engineering and the rise of GitHub Copilot

Programming has always been a mixture of science, engineering, and creativity. Creating new programs and being able to instruct computers to do something has always been something that was considered worth paying for – that's how all programmers make their living. There have been attempts to automate programming and to support smaller tasks – for example, provide programmers with suggestions on how to use a specific function or library method.

Good programmers, however, can make programs that last and that are readable for others. They can also make reliable programs that work without maintenance for a long period. The best programmers are the ones who can solve very difficult tasks and follow the principles and best practices of software engineering.

In 2020, something happened – GitHub Copilot entered the stage and showed that automated tools, based on **large language models** (**LLMs**), can provide much more than just suggestions for simple function calls. It has demonstrated that these language models are capable of providing suggestions for entire solutions and algorithms and even solving programming competitions. This opened up completely new possibilities for programmers – the best ones became extremely productive, and have been provided with the tools that allow them to focus on the complex parts of their programming tasks. The simple ones are now solved by GitHub Copilot and others.

The reason why these tools are so good is because they are based on LLMs, which are capable of finding and quantifying contexts of programs. Just like a great chess player can foresee several moves in advance, these tools can foresee what the programmers may need in advance and provide useful suggestions.

There are a few simple tricks that make these tools so effective, and one of them is feature engineering. Feature engineering for natural language tasks, including programming, is a process where a piece of text is transformed into a vector (or matrix) of numbers. These vectors can be simple – for example, quantifying the tokens – and also very complex – for example, finding an atomic piece of text linked to other tasks. We'll explore these techniques in this chapter. We'll start with a bit of a repetition of the bag-of-words technique (seen in *Chapter 3* and *Chapter 5*). We do not need to repeat the entire code, but we do need to provide a small re-cap to understand the limitations of these approaches. However, here is my best practice for choosing whether I need a tokenizer or embeddings.

> **Best practice #44**
> Use tokenizers for LLMs such as BERT and word embeddings for simple tasks.

For simple tasks, such as basic tokenization of the text for sentiment analysis or quickly understanding the dependencies in the text, I often use word embeddings. However, I usually use different tokenizers for working with LLMs such as BERT, RoBERTa, or AlBERT since these models are very good at finding dependencies on their own. However, for designing classifiers, I use word embeddings since they provide a fast way of creating feature vectors that are compatible with the "classical" machine learning algorithms.

Choosing a tokenizer needs to be done based on the task. We'll look at this closer in this chapter, but the topic itself could occupy an entire book. For example, for tasks that require information about the part of speech (or, in many cases, part of the abstract syntax tree of a program), we need to use a tokenizer that is designed to capture that information – for example, from a programming language parser. These tokenizers provide more information to the model, but they impose more requirements on the data – an abstract syntax tree-based tokenizer requires the program to be well formed.

What a tokenizer is and what it does

The first step in feature engineering text data is to decide on the tokenization of the text. The tokenization of text is a process of extracting parts of words that capture the meaning of the text without too many extra details.

There are different ways to extract tokens, which we'll explore in this chapter, but to illustrate the problem of extracting tokens, let's look at one word that can take different forms – *print*. The word by itself can be a token, but it can be in different forms, such as *printing, printed, printer, prints, imprinted,* and many others. If we use a simple tokenizer, each of these words will be one token – which means quite a few tokens. However, all these tokens capture some sort of meaning related to printing, so maybe we do not need so many of them.

This is where tokenizers come in. Here, we can decide how to treat these different forms of the word. We could take the main part only – *print* – and then all the other forms would be counted as that, so both *imprinted* and *printing* would be counted as *print*. It decreases the number of tokens, but we reduce the expressiveness of our feature vector – some information is lost as we do not have the same number of tokens to use. We could pre-design the set of tokens – that is, use both *print* and *imprint* to distinguish between different contexts. We could also use bigrams (two words together) as tokens (for example, *is_going* versus is, *going* – the first one requires both words to be in the specific sequence, where the other one allows them to be in two different sequences), or we could add the information about whether the word is an object of the subject in a sentence.

Bag-of-words and simple tokenizers

In *Chapters 3* and *5*, we saw the use of the bag-of-words feature extraction technique. This technique takes the text and counts the number of tokens, which were words in *Chapters 3* and *5*. It is simple and computationally efficient, but it has a few problems.

When instantiating the bag-of-words tokenizer, we can use several parameters that strongly impact the results, as we did in the following fragment of code in the previous chapters:

```
# create the feature extractor, i.e., BOW vectorizer
# please note the argument - max_features
# this argument says that we only want three features
# this will illustrate that we can get problems - e.g. noise
# when using too few features
vectorizer = CountVectorizer(max_features = 3)
```

The `max_features` parameter is a cut-off value that reduces the number of features, but it also can introduce noise where two (or more) distinct sentences have the same feature vector (we saw an example of such a sentence in *Chapter 2*). Since we discussed noise and the problems related to it, we could be tempted to use other parameters – `max_df` and `min_df`. These two parameters determine how often a word should appear in the document to be considered a token. The tokens that are too rare can (`min_df`) result in a sparse matrix – a lot of 0s in the feature matrix – but they can be a very good discriminant between data points. Maybe these rare words are just what we are looking for. The other parameter (`max_df`) results in more dense feature matrices, but they may not discriminate the data points completely. This means that it is not so simple to select these parameters – we need experiments and we need to use machine learning model training (and validation) to find the right vector.

There is also another way – we can perform a recursive search for such a feature vector that would discriminate all data points without adding too much noise. My team has experimented with such algorithms, which yield very good performance for model training and validation but are computationally very expensive. Such an algorithm is presented in *Figure 8.1*:

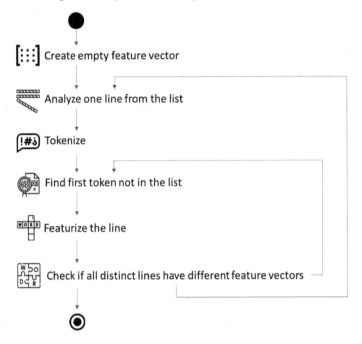

Figure 8.1 – An algorithm for finding a set of features that discriminate all data points
in a text file. The flow has been simplified to illustrate the main points

The algorithm works by adding new tokens if a data point has the same feature vector as any of the previous ones. It starts by taking the first token from the first line, then the second line. If the token can discriminate between these two lines, then it proceeds to the third line. Once the algorithm discovers that two different lines have the same feature vector, it finds out whether there is a token that can discriminate between these lines and adds it to the set of features. It continues until there are no new tokens to add or all lines have been analyzed.

This algorithm guarantees that the set of tokens that best discriminates the analyzed dataset is found. However, it has one large disadvantage – it is slow (as it must start from the first line once a new token is found/needed). The resulting feature matrix is also not optimal – it contains a lot of 0s since most of the tokens can only be found in one line. The feature matrix, in turn, can be much larger than the actual raw dataset.

This is where my next best practice comes in.

> **Best practice #45**
>
> Use bag-of-words tokenizers together with dictionaries when your task requires a pre-defined set of words.

I use bag-of-words tokenizers quite often when analyzing programming language code. I use a pre-defined set of keywords from the programming language to boost the tokenizer and then the standard `CountVectorizer`. This allows me to control part of the vocabulary that I am interested in – keywords – and allows the tokenizer to adjust to the text.

WordPiece tokenizer

A better way to tokenize and extract features from text documents is to use a WordPiece tokenizer. This tokenizer works in such a way that it finds the most common pieces of text that it can discriminate, and also the ones that are the most common. This kind of tokenizer needs to be trained – that is, we need to provide a set of representative texts to get the right vocabulary (tokens).

Let's look at an example where we use a simple program, a module from an open source project, to train such a tokenizer and then apply this tokenizer to the famous "Hello World" program in C. Let's start by creating the tokenizer:

```
from tokenizers import BertWordPieceTokenizer

# initialize the actual tokenizer
tokenizer = BertWordPieceTokenizer(
    clean_text=True,
    handle_chinese_chars=False,
    strip_accents=False,
    lowercase=True
)
```

In this example, we're using the WordPiece tokenizer from the Hugging Face library, specifically the one that is prepared to work with LLMs such as BERT. There are several parameters that we can use, but let's settle with the parameters that show that we are only interested in lowercase characters; we do not want to handle Chinese characters and want to start from scratch.

Now, we need to find a piece of text that we can train the tokenizer on. In this example, I'll use one of the files from an open source project – AzureOS NetX. It's a component written in C that handles parts of the internet HTTP protocol. We create a new variable – `path` – and add the path to that file there. Once we've prepared the text, we can train the tokenizer:

```
# and train the tokenizer based on the text
tokenizer.train(files=paths,
                vocab_size=30_000,
```

```
                    min_frequency=1,
                    limit_alphabet=1000,
                    wordpieces_prefix='##',
                    special_tokens=['[PAD]', '[UNK]', '[CLS]', '[SEP]',
'[MASK]'])
```

We've set the tokenizer to a similar set of parameters, similar to `CountVectorizer` in the previous examples. This preceding code fragment finds the set of the most common pieces of words and uses them as tokens.

We can get the list of tokens through the `tokenizer.get_vocab()` statement, which results in a long dictionary of tokens. Here are the first few ones:

```
'##ll': 183,
'disable': 326,
'al': 263,
'##cket': 90,
'##s': 65,
'computed': 484
```

The first token is a piece of a word, which is denoted by the fact that it has two hashtags at the beginning of it. This token is mapped to the number `183` in the vocabulary. This mapping is important as the numbers are used later on by the machine learning models.

Another interesting observation is that some of the tokens, such as `'disable'`, are not pieces of words but entire words. This means that this token does not appear as a piece of the word anywhere and it does not contain any other pieces of other words in the vocabulary.

Once we've trained the WordPiece tokenizer, we can check how the tokenizer extracts features from a simple C program:

```
strCProgram = '''
int main(int argc, void **argc)
{
  printf("%s", "Hello World\n");
  return 0;
}
'''


# now, let's see how the tokenizer works
# we invoke it based on the program above
tokenizedText = tokenizer.encode(strCProgram)

tokenizedText.tokens
```

The preceding piece of code tokenizes the program. The result is the following list of tokens (only the first 10 tokens of 50 are shown):

```
'in', '##t', 'ma', '##in', '(', 'in', '##t', 'a', '##r', '##g'
```

The first line, which starts with the int token, has been tokenized in the following way. The first word – int – is split into two tokens: "in" and "##t". This is because these two parts were used in the training program. We can also see that the second token – main – is split into two tokens: "ma" and "##in". The IDs for these tokens are as follows:

```
110, 57, 272, 104, 10, 110, 57, 30, 61, 63
```

This means that this list of numbers is the feature vector for our simple C program.

WordPiece tokenization is very effective, but it depends a lot on the training data. If we use training data that is very different from the tokenized text, the set of tokens will not be very helpful. Therefore, my next best practice is about training this tokenizer.

> **Best practice #46**
> Use the WordPiece tokenizer as your first choice.

I usually use this tokenizer as my first choice. It is relatively flexible but quite fast. It allows us to capture a vocabulary that does the job most of the time and does not require a lot of setup. For simple tasks with straightforward language and a well-defined vocabulary, traditional word-level tokenization or other subword tokenization methods such as **byte-pair encoding** (**BPE**) may suffice. WordPiece tokenization can increase the size of the input data due to the introduction of subword tokens. This can impact memory and computational requirements.

BPE

A more advanced method for tokenizing text is the BPE algorithm. This algorithm is based on the same premises as the compression algorithm that was created in the 1990s by Gage. The algorithm compresses a series of bytes by the bytes not used in the compressed data. The BPE tokenizer does a similar thing, except that it replaces a series of tokens with new bytes that are not used in the text. In this way, the algorithm can create a much larger vocabulary than CountVectorizer and the WordPiece tokenizer. BPE is very popular both for its ability to handle large vocabulary and for its efficient implementation through the fastBPE library.

Let's explore how to apply this tokenizer to the same data and check the difference between the previous two. The following code fragment shows how to instantiate this tokenizer from the Hugging Face library:

```
# in this example we use the tokenizers
# from the HuggingFace library
from tokenizers import Tokenizer
```

```
from tokenizers.models import BPE

# we instantiate the tokenizer
tokenizer = Tokenizer(BPE(unk_token="[UNK]"))
```

This tokenizer requires training as it needs to find the optimal set of pairs of tokens. Therefore, we need to instantiate a trainer class and train it. The following piece of code does just that:

```
from tokenizers.trainers import BpeTrainer

# here we instantiate the trainer
# which is a specific class that will manage
# the training process of the tokenizer
trainer = BpeTrainer(special_tokens=["[UNK]", "[CLS]",
                     "[SEP]", "[PAD]", "[MASK]"])


from tokenizers.pre_tokenizers import Whitespace

tokenizer.pre_tokenizer = Whitespace()

# now, we need to prepare a dataset
# in our case, let's just read a dataset that is a code of a program

# in this example, I use the file from an open-source component -
Azure NetX
# the actual part is not that important, as long as we have a set of
# tokens that we want to analyze
paths = ['/content/drive/MyDrive/ds/cs_dos/nx_icmp_checksum_
compute.c']

# finally, we are ready to train the tokenizer
tokenizer.train(paths, trainer)
```

The important part of this training process is the use of a special pre-tokenizer. The pre-tokenizer is how we initially split words into tokens. In our case, we use the standard whitespaces, but we could use something more advanced. For example, we could use semicolons and therefore use entire lines of code as tokens.

After executing the preceding fragment of code, our tokenizer is trained and ready to use. We can check the tokens by writing `tokenizer.get_vocab()`. The set of tokens is as follows (the first 10 tokens):

```
'only': 565, 'he': 87, 'RTOS': 416, 'DE': 266, 'CH': 154, 'a': 54,
'ps': 534, 'will': 372, 'NX_SHIFT_BY': 311, 'O': 42,
```

This set of tokens is very different from the set of tokens in previous cases. It contains a mix of words such as "will" and subwords such as "ol." This is because the BPE tokenizer found some replicated tokens and replaced them with dedicated bytes.

> **Best practice #47**
> Use BPE when working with LLMs and large corpora of text.

I use BPE as my go-to when I analyze large pieces of text, such as large code bases. It is blazingly fast for this task and can capture complex dependencies. It is also heavily used in models such as BERT or GPT.

Now, in our case, the program's source code that we used to train the BPE tokenizer was small, so a lot of words did not repeat themselves and the optimization does not make much sense. Therefore, a WordPiece tokenizer would do an equally (if not better) job. However, for larger text corpora, this tokenizer is much more effective and efficient than WordPiece or bag-of-words. It is also the basis for the next tokenizer – SentencePiece.

The SentencePiece tokenizer

SentencePiece is a more general option than BPE for one more reason: it allows us to treat whitespaces as regular tokens. This allows us to find more complex dependencies and therefore train models that understand more than just pieces of words. Hence the name – SentencePiece. This tokenizer was originally introduced to enable the tokenization of languages such as Japanese, which do not use whitespaces in the same way as, for example, English. The tokenizer can be installed by running the `pip install -q sentencepiece` command.

In the following code example, we're instantiating and training the SentencePiece tokenizer:

```
import sentencepiece as spm

# this statement trains the tokenizer
spm.SentencePieceTrainer.train('--input="/content/drive/MyDrive/ds/
cs_dos/nx_icmp_checksum_compute.c" --model_prefix=m --vocab_size=200')

# makes segmenter instance and
# loads the model file (m.model)
sp = spm.SentencePieceProcessor()
sp.load('m.model')
```

We've trained it on the same file as the previous tokenizers. The text was a programming file, so we could expect the tokenizer to give us a better understanding of what's in a programming language than what's in a normal piece of text. Something worth noting is the size of the vocabulary, which is 200, unlike 30,000 in the previous examples. This is important because this tokenizer tries to find as many tokens as this parameter. Since our input program is very short – one file with a few functions in it – the tokenizer cannot create more than about 300 tokens.

The following fragment encodes the "Hello World" program using this tokenizer and prints the following output:

```
strCProgram = '''
int main(int argc, void **argc)
{
  printf("%s", "Hello World\n");
  return 0;
}
'''
print(sp.encode_as_pieces(strCProgram))
```

The first 10 tokens are represented in the following way:

```
'_in', 't', '_', 'm', 'a', 'in', '(', 'in', 't', '_a'
```

The new element in this tokenizer is the underscore character (_). It denotes whitespace in the text. This is unique and it allows us to use this tokenizer more effectively in programming language comprehension because it allows us to capture such programming constructs as nesting – that is, using tabs instead of spaces or writing several statements in the same line. This is all because this tokenizer treats whitespaces as something important.

> **Best practice #48**
> Use the SentencePiece tokenizer when no clear word boundaries are present.

I use SentencePiece when analyzing programming language code with a focus on programming styles – for example, when we focus on things such as camel-case variable naming. For this task, it is important to understand how programmers use spaces, formatting, and other compiler-transparent elements. Therefore, this tokenizer is perfect for such tasks.

Word embeddings

Tokenizers are one way of extracting features from text. They are powerful and can be trained to create complex tokens and capture statistical dependencies of words. However, they are limited by the fact that they are completely unsupervised and do not capture any meaning or relationship between words. This means that the tokenizers are great at providing input to neural network models, such as BERT, but sometimes, we would like to have features that are more aligned with a certain task.

This is where word embeddings come to the rescue. The following code shows how to instantiate the word embedding model, which is imported from the gensim library. First, we need to prepare the dataset:

```
from gensim.models import word2vec
# now, we need to prepare a dataset
# in our case, let's just read a dataset that is a code of a program

# in this example, I use the file from an open source component -
Azure NetX
# the actual part is not that important, as long as we have a set of
# tokens that we want to analyze
path = '/content/drive/MyDrive/ds/cs_dos/nx_icmp_checksum_compute.c'


# read all lines into an array
with open(path, 'r') as r:
  lines = r.readlines()

# and see how many lines we got
print(f'The file (and thus our corpus) contains {len(lines)} lines')
```

The preceding code fragment prepares the file differently compared to the tokenizers. It creates a list of lines, and each line is a list of tokens, separated by whitespaces. Now, we are ready to create the word2vec model and train it on this data:

```
# we need to pass splitted sentences to the model
tokenized_sentences = [sentence.split() for sentence in lines]


model = word2vec.Word2Vec(tokenized_sentences,
                          vector_size=10,
                          window=1,
                          min_count=0,
                          workers=4)
```

The result is that the model is trained on the corpus that we provided – the C program implementing a part of the HTTP protocol. We can look at the first 10 tokens that have been extracted by writing model.wv.key_to_index:

```
'*/': 0, '/*': 1, 'the': 2, '=': 3, 'checksum': 4, '->': 5, 'packet':
6, 'if': 7, 'of': 8, '/*********************************************
**********************/': 9,
```

In total, word2vec extracted 259 tokens.

This word embedding model is different from the tokenizers that we used before. It embeds the values of the words (tokens) into a latent space, which allows us to utilize the lexical properties of these words more smartly. For example, we can check the similarity of words using `model.wv.most_similar(positive=['add'])`:

```
('NX_LOWER_16_MASK;', 0.8372778296470642),
('Mask', 0.8019374012947083),
('DESCRIPTION', 0.7171915173530579),
```

We can also pretend that these words are vectors and their similarity is captured in this vector. Therefore, we can write something like `model.wv.most_similar(positive=['file', 'function'], negative=['found'])` and obtain a result like this:

```
('again', 0.24998697638511658),
('word', 0.21356187760829926),
('05-19-2020', 0.21174617111682892),
('*current_packet;', 0.2079058289527893),
```

The expression will be similar if we use mathematics to express it: *result = file + function – found*. The resulting list of similar words is the list of words that are the closest to the vector that was captured as the result of this calculation.

Word embeddings are very powerful when we want to capture the similarity of the words and expressions. However, the original implementation of this model has certain limitations – for example, it does not allow us to use words that are not part of the original vocabulary. Asking for a word that is similar to an unknown token (for example, `model.wv.most_similar(positive=['return'])`) results in an error.

FastText

Luckily for us, there is an extension of the `word2vec` model that can approximate the unknown tokens – FastText. We can use it in a very similar way as we use `word2vec`:

```
from gensim.models import FastText

# create the instance of the model
model = FastText(vector_size=4,
                 window=3,
                 min_count=1)

# build a vocabulary
model.build_vocab(corpus_iterable=tokenized_sentences)

# and train the model
model.train(corpus_iterable=tokenized_sentences,
```

```
                    total_examples=len(tokenized_sentences),
                    epochs=10)
```

In the preceding code fragment, the model is trained on the same set of data as word2vec. model = FastText(vector_size=4, window=3, min_count=1) creates an instance of the FastText model with three hyperparameters:

- vector_size: The number of elements in the resulting feature vector

- window: The size of the window used to capture context words

- min_count: The minimum frequency of a word to be included in the vocabulary

model.build_vocab(corpus_iterable=tokenized_sentences) builds the vocabulary of the model by iterating through the tokenized_sentences iterable (which should contain a list of lists, with each inner list representing a sentence tokenized into individual words) and adding each word to the vocabulary if it meets the min_count threshold. model.train(corpus_iterable=tokenized_sentences, total_examples=len(tokenized_sentences), epochs=10) trains the FastText model using the tokenized_sentences iterable for a total of 10 epochs. During each epoch, the model iterates through the corpus again and updates its internal weights based on the context words surrounding each target word. The total_examples parameter tells the model how many total examples (that is, sentences) are in the corpus, which is used to calculate the learning rate.

The input is the same. However, if we invoke the similarity for the unknown token, such as model.wv.most_similar(positive=['return']), we get the following result:

```
('void', 0.5913326740264893),
('int', 0.43626993894577026),
('{', 0.2602742612361908),
```

The set of three similar words indicates that the model can approximate an unknown token.

My next best practice is about the use of FastText.

> **Best practice #49**
>
> Use word embeddings, such as FastText, as a valuable feature representation for text classification tasks, but consider incorporating them into more comprehensive models for optimal performance.

Unless we need to use an LLM, this kind of feature extraction is a great alternative to the simple bag-of-words technique and powerful LLMs. It captures some parts of the meaning and allows us to design classifiers based on text data. It can also handle unknown tokens, which makes it very flexible.

From feature extraction to models

The feature extraction methods presented in this chapter are not the only ones we can use. Quite a few more exist (to say the least). However, they all work similarly. Unfortunately, no silver bullet exists, and all models have advantages and disadvantages. For the same task, but a different dataset, simpler models may be better than complex ones.

Now that we have seen how to extract features from text, images, and numerical data, it's time we start training the models. This is what we'll do in the next chapter.

References

- *Al-Sabbagh, K.W., et al. Selective regression testing based on big data: comparing feature extraction techniques. in 2020 IEEE International Conference on Software Testing, Verification and Validation Workshops (ICSTW). 2020. IEEE.*

- *Staron, M., et al. Improving Quality of Code Review Datasets–Token-Based Feature Extraction Method. in Software Quality: Future Perspectives on Software Engineering Quality: 13th International Conference, SWQD 2021, Vienna, Austria, January 19–21, 2021, Proceedings 13. 2021. Springer.*

- *Sennrich, R., B. Haddow, and A. Birch, Neural machine translation of rare words with subword units. arXiv preprint arXiv:1508.07909, 2015.*

- *Gage, P., A new algorithm for data compression. C Users Journal, 1994. 12(2): p. 23-38.*

- *Kudo, T. and J. Richardson, SentencePiece: A simple and language independent subword tokenizer and detokenizer for neural text processing. arXiv preprint arXiv:1808.06226, 2018.*

Part 3:
Design and Development
of ML Systems

Although machine learning and its cousin artificial intelligence are well known, they refer to a wide range of algorithms and models. First, there are the classical machine learning models that are based on statistical learning and usually require the data to be prepared in a tabular form. They identify patterns in the data and can replicate these patterns. However, there are also modern models that are based on deep learning, which are capable of capturing more fine-grained patterns in data that is less structured. The crown examples of these models are the transformer models (GPT) and autoencoders (diffusers). In this part of the book, we take a closer look at these models. We focus on how these models can be trained and integrated into machine learning pipelines. We also look into how software engineering practices should be applied to these models.

This part has the following chapters:

- *Chapter 9, Types of Machine Learning Systems – Feature-Based and Raw Data-Based (Deep Learning)*

- *Chapter 10, Training and Evaluation of Classical ML Systems and Neural Networks*

- *Chapter 11, Training and Evaluation of Advanced ML Algorithms – GPT-3 and Autoencoders*

- *Chapter 12, Designing Machine Learning Pipelines and Their Testing*

- *Chapter 13, Designing and Implementation of Large-Scale, Robust ML Software*

9

Types of Machine Learning Systems – Feature-Based and Raw Data-Based (Deep Learning)

In the previous chapters, we learned about data, noise, features, and visualization. Now, it's time to move on to machine learning models. There is no such thing as one model, but there are plenty of them – starting from the classical models such as random forest to deep learning models for vision systems to generative AI models such as GPT.

The convolutional and GPT models are called deep learning models. Their name comes from the fact that they use raw data as input and the first layers of the models include feature extraction layers. They are also designed to progressively learn more abstract features as the input data moves through these models.

This chapter demonstrates each of these types of models and progresses from classical machine learning to generative AI models.

In this chapter, we'll cover the following topics:

- Why do we need different types of models?
- Classical machine learning models and systems, such as random forest, decision tree, and logistic regression
- Deep learning models for vision systems, convolutional neural models, and **You Only Look Once (YOLO)** models
- **General Pretrained Transformers (GPT)** models

Why do we need different types of models?

So far, we have invested a significant amount of effort in data processing while focusing on tasks such as noise reduction and annotation. However, we have yet to delve into the models that are employed to work with this processed data. While we briefly mentioned different types of models based on data annotation, including supervised, unsupervised, and reinforced learning, we have not thoroughly explored the user's perspective when it comes to utilizing these models.

It is important to consider the perspective of the user when employing machine learning models for working with data. The user's needs, preferences, and specific requirements play a crucial role in selecting and utilizing the appropriate models.

From the user's standpoint, it becomes essential to assess factors such as model interpretability, ease of integration, computational efficiency, and scalability. Depending on the application and use case, the user might prioritize different aspects of the models, such as accuracy, speed, or the ability to handle large-scale datasets.

Furthermore, the user's domain expertise and familiarity with the underlying algorithms impact the selection and evaluation of models. Some users might prefer simpler, more transparent models that offer interpretability and comprehensibility, while others might be willing to trade interpretability for improved predictive performance using more complex models such as deep learning networks.

Considering the user's perspective enables a more holistic approach to model selection and deployment. It involves actively involving the user in the decision-making process, gathering feedback, and continuously refining the models to meet their specific needs.

By incorporating the user's perspective into the discussion, we can ensure that the models we choose not only satisfy technical requirements but also align with the user's expectations and objectives, ultimately enhancing the effectiveness and usability of the entire system.

Therefore, moving forward, we'll explore how different types of users interact with and benefit from various machine learning models while considering their specific requirements, preferences, and domain expertise. We'll start with the classical machine learning models, which are historically the first ones.

Classical machine learning models

Classical machine learning models require pre-processed data in the form of tables and matrices. Classical machine learning models, such as random forest, linear regression, and support vector machines, require a clear set of predictors and classes to find patterns. Due to this, our pre-processing pipelines need to be manually designed for the task at hand.

From the user's perspective, these systems are designed in a very classical way – there is a user interface, an engine for data processing (our classical machine learning model), and an output. This is depicted in *Figure 9.1*:

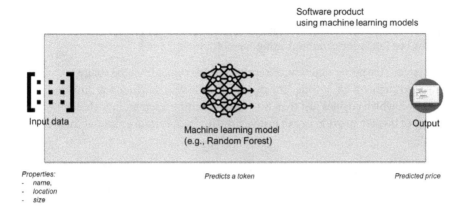

Figure 9.1 – Elements of a machine learning system

Figure 9.1 shows that there are three elements – the input prompt, the model, and the output. For most such systems, the input prompt is a set of properties that are provided for the model. The user fills in some sort of form and the system provides an answer. It can be a form for predicting the price of land or a system for loans, applying for a job, finding the best car, and so on.

The source code for such a system may look something like this:

```
import pandas as pd
from sklearn.linear_model import LinearRegression
from sklearn.model_selection import train_test_split
from sklearn.metrics import mean_squared_error


# Load the stock price data into a pandas DataFrame
data = pd.read_csv('land_property_data.csv')


# Select the features (e.g., historical prices, volume, etc.) and the
target variable (price)
X = data[['Feature1', 'Feature2', ...]]  # Relevant features here
y = data['price']


# read the model from the serialized storage here


# Make predictions on the test data
y_pred = model.predict(X)


# Evaluate the model using mean squared error (MSE)
print(f'The predicted value of the property is: {y_pred}')
```

This fragment of code requires a model to be already trained and only uses it for making predictions. The main line that uses the model is the line in boldface. The rest of the code fragment is for processing the input and the last line is for communicating the output.

In modern ecosystems, the power of machine learning models comes from the ability to change models without the need to change a lot of code. The majority of classical machine learning models use this fit/predict interface, which enables just that. So, which machine learning models can we use? There are just too many of them to provide an exhaustive list. However, certain groups of these models have certain properties:

- **Regression models** group machine learning models that are used for predicting a class value. They can be used both for classification (classifying a module to be defect-prone or not) and prediction tasks (predicting the number of defects in a module). These models are based on finding the best curve to fit the given data.

- **Tree-based models** group models that are based on finding differences in the dataset as if we wrote a set of if-then statements. The logical conditions for these if-then statements are based on the statistical properties of the data. These models are good for both classification and prediction models.

- **Clustering algorithms** group models that are based on finding similarities in the data and grouping similar entities. They are often unsupervised and require some experimentation to find the right set of parameters (for example, the number of clusters).

- **Neural networks** group all kinds of neural networks that can be used for classical machine learning tasks. These algorithms require us to design and train the neural network model.

We can select these models based on their properties and test them to find the best one. However, if we include hyperparameter training, this process is very time-consuming and effort-intensive. Therefore, I strongly recommend using AutoML approaches for this. AutoML is a group of algorithms that utilize the fit/predict interface for machine learning models to find the best model automatically. By exploring the plethora of models, they can find the model that is the best for the dataset. We say this is with an asterisk. Sometimes, the human ability to understand the data and its properties beats most automated machine learning processes (`https://metrics.blogg.gu.se/?p=682`).

So, here is my first best practice for this chapter.

> **Best practice #50**
>
> Use AutoML as your first choice when you're training classical machine learning models.

Using AutoML is very simple and can be illustrated with the following fragment of code (from the documentation of auto-sklearn):

```
import autosklearn.classification
cls = autosklearn.classification.AutoSklearnClassifier()
```

```
cls.fit(X_train, y_train)
predictions = cls.predict(X_test)
```

The preceding fragment illustrates how easy it is to use the auto-sklearn toolkit to find the best model. Please note that this toolkit has been designed for Linux-based systems only. To use it on the Microsoft Windows operating system, I recommend using **Windows Subsystem for Linux 2.0** (**WSL 2**). The interface hides the best model in such a way that the user does not even have to see which model is the best for the data at hand.

`import autosklearn.classification` imports the auto-sklearn module specifically for classification tasks. `cls = autosklearn.classification.AutoSklearnClassifier()` initializes an instance of the `AutoSklearnClassifier` class, which represents the AutoML classifier in `autosklearn`. It creates an object that will be used to search for the best classifier and its hyperparameters automatically. `cls.fit(X_train, y_train)` fits `AutoSklearnClassifier` to the training data. It automatically explores different classifiers and their hyperparameter configurations to find the best model based on the provided `X_train` (features) and `y_train` (target labels). It trains the AutoML model on the provided training dataset.

`predictions = cls.predict(X_test)` uses the fitted `AutoSklearnClassifier` to make predictions on the `X_test` dataset. It applies the best-found model from the previous step to the test data and assigns the predicted labels to the `predictions` variable.

Let's apply auto-sklearn on the same dataset that we used for visualization in *Chapter 6*:

```
# read the file with data using openpyxl
import pandas as pd

# we read the data from the excel file,
# which is the defect data from the ant 1.3 system
dfDataCamel12 = pd.read_excel('./chapter_6_dataset_numerical.xlsx',
                              sheet_name='camel_1_2',
                              index_col=0)
# prepare the dataset
import sklearn.model_selection

X = dfDataCamel12.drop(['Defect'], axis=1)
y = dfDataCamel12.Defect

X_train, X_test, y_train, y_test = \
        sklearn.model_selection.train_test_split(X, y, random_
state=42, train_size=0.9)
```

We'll use the same code we used previously:

```
import autosklearn.classification
cls = autosklearn.classification.AutoSklearnClassifier()
```

```
cls.fit(X_train, y_train)
predictions = cls.predict(X_test)
```

Once we have trained the model, we can inspect it – for example, by asking auto-sklearn to provide us with information about the best model – using the `print(cls.sprint_statistics())` command. The results are as follows:

```
auto-sklearn results:
Dataset name: 4b131006-f653-11ed-814a-00155de31e8a
Metric: accuracy
Best validation score: 0.790909
Number of target algorithm runs: 1273
Number of successful target algorithm runs: 1214
Number of crashed target algorithm runs: 59
Number of target algorithms that exceeded the time limit: 0
Number of target algorithms that exceeded the memory limit: 0
```

This information shows us that the toolkit has tested 1273 algorithms and that 59 of them crashed. This means that they were not compatible with the dataset provided by us.

We can also ask the toolkit to provide us with the best model by using the `print(cls.show_models())` command. This command provides a long list of the models that are used for ensemble learning and their weight on the final score. Finally, we can ask for the accuracy score for the test data by using `print(f\"Accuracy score {sklearn.metrics.accuracy_score(y_test, predictions):.2f}\")`. For this dataset, the accuracy score is 0.59 for the test data, which is not a lot. However, this is the model that's obtained by using the best ensemble. If we ask the model to provide us with the accuracy score for the training data, we'll get 0.79, which is much higher, but that's because the model is very well optimized.

Later in this book, we'll explore these algorithms and learn how they behave for tasks in software engineering and beyond.

Convolutional neural networks and image processing

The classical machine learning models are quite powerful, but they are limited in their input. We need to pre-process it so that it's a set of feature vectors. They are also limited in their ability to learn – they are one-shot learners. We can only train them once and we cannot add more training. If more training is required, we need to train these models from the very beginning.

The classical machine learning models are also considered to be rather limited in their ability to handle complex structures, such as images. Images, as we have learned before, have at least two different dimensions and they can have three channels of information – red, green, and blue. In more complex applications, the images can contain data from LiDAR or geospatial data that can provide meta-information about the images.

So, to handle images, more complex models are needed. One of these models is the YOLO model. It's considered to be state-of-the-art in the area of object detection due to its great balance between accuracy and speed.

Let's take a look at how we can utilize a pre-trained YOLO v5 model from Hugging Face. Here, I would like to provide my next best practice.

> **Best practice #51**
> Use pre-trained models from Hugging Face or TensorFlow Hub to start with.

Using a pre-trained model has a few advantages:

- First of all, it allows us to use the network as a benchmark for our pipeline. We can experiment with it and understand its limitations before we move forward and start training it.

- Second, it provides us with the possibility to add more training for the existing, proven-in-use models that others have also used.

- Finally, it provides us with the possibility to share our models with the community to support the ethical and responsible development of artificial intelligence.

The following code fragment installs the YoLo model and instantiates it:

```
# install YoLo v5 network
!pip install -q -U yolov5
# then we set up the network
import yolov5

# load model
model = yolov5.load('fcakyon/yolov5s-v7.0')

# set model parameters
model.conf = 0.25  # NMS confidence threshold
model.iou = 0.45  # NMS IoU threshold
model.agnostic = False  # NMS class-agnostic
model.multi_label = False  # NMS multiple labels per box
model.max_det = 1000  # maximum number of detections per image
```

The first few lines load the YOLOv5 model from the specified source – that is, fcakyon/yolov5s-v7.0 – using the load function. They assign the loaded model to the variable model, which can be used to perform object detection. The model.conf parameter sets the confidence threshold for **non-maximum suppression** (**NMS**), which is used to filter out detections below this confidence level. In this case, it is set to 0.25, meaning that only detections with a confidence score above 0.25 will be considered.

The model.iou parameter sets the **Intersection over Union** (**IoU**) threshold for NMS. It determines the degree of overlap between bounding boxes required to consider them as duplicate detections. Here, it is set to 0.45, meaning that if the IoU between two boxes is above 0.45, the one with the lower confidence score will be suppressed. The model.agnostic parameter determines whether NMS is class-agnostic or not. If it's set to False, NMS will consider class labels during suppression, which means that if two bounding boxes have the same coordinates but different labels, they will not be considered duplicates. Here, it is set to False. The model.multi_label parameter controls whether NMS allows multiple labels per bounding box. If it's set to False, each box will be assigned a single label with the highest confidence score. Here, it is set to False.

Finally, the model.max_det parameter sets the maximum number of detections allowed per image. In this case, it is set to 1000, meaning that only the top 1,000 detections (sorted by confidence score) will be kept.

Now, we can perform inferences – that is, detect objects using the network – but first, we must load the image:

```
# and now we prepare the image
from PIL import Image
from torchvision import transforms


# Load and preprocess the image
image = Image.open('./test_image.jpg')
```

This code fragment loads the image file located at ./test_image.jpg using the open function from PIL's Image module. It creates an instance of the Image class representing the image.

Once the image has been loaded, you can apply various transformations to pre-process it before feeding it to the YOLOv5 model for object detection. This might involve resizing, normalization, or other pre-processing steps, depending on the model's requirements:

```
# perform inference
results = model(image)

# inference with larger input size
results = model(image, size=640)

# inference with test time augmentation
results = model(image, augment=True)

# parse results
predictions = results.pred[0]
boxes = predictions[:, :4] # x1, y1, x2, y2
scores = predictions[:, 4]
categories = predictions[:, 5]
```

```
# show detection bounding boxes on image
results.show()
```

The preceding code fragment performs object detection in the first few lines and then draws the image, together with the bounding boxes of the detected object. In our case, this is the result of the preceding code fragment:

Figure 9.2 – Objects detected in the image

Please note that the model identifies the car as a truck, perhaps because of the presence of the additional luggage on the back of the car. The source of the image is Pixabay. The figure shows that object detection does not identify the object correctly. However, this is not a problem. We can take this pre-trained model and train it even more. However, that is the topic of the next few chapters, so we won't cover it here.

> **Best practice #52**
>
> Work with pre-trained networks to identify their limitations and then train the network on your own dataset.

I strongly recommend using the pre-trained models to start with and then train the network on your own data. This ability of deep learning models to continue training is a great property that we can utilize when designing machine learning-based systems. In this way, we get the best of both worlds – our systems can detect generic objects while being better at detecting objects that our system specifies. This kind of approach is often used in designing automotive systems.

Let's look at other types of deep learning systems that use pre-training and add one more layer of complexity – prompt engineering.

BERT and GPT models

BERT and GPT models use raw data as input and their main output is one predicted word. This word can be predicted both in the middle of a sentence and at the end of it. This means that the products that are designed around these models need to process data differently than in the other models.

Figure 9.3 provides an overview of this kind of processing with a focus on both prompt engineering in the beginning and output processing in the end. This figure shows the machine learning models based on the BERT or GPT architecture in the center. This is an important aspect, but it only provides a very small element of the entire system (or tool).

The tool's workflow starts on the left-hand side with input processing. For the user, it is a prompt that asks the model to do something, such as `"Write a function that reverses a string in C"`. The tool turns that prompt into a useful input for the model – it can find a similar C program as input for the model and add the `<mask>` token to the end of that program to make the model predict the output.

Then, the model makes the prediction – one token at a time. The tool needs to loop the output until something useful is generated. The tool needs to define the stop criteria and it needs to process the output into a useful one for the user:

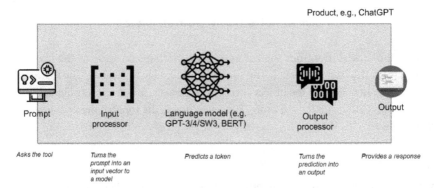

Figure 9.3 – Overview of the design of modern deep learning tools where the machine learning model is only a small (but important) part of the tool

If we send this prompt to ChatGPT, we get the following response in three parts. First, we get the actual code of the function, which reverses the string in C:

```
#include <stdio.h>
#include <string.h>

void reverseString(char* str) {
```

```
    int length = strlen(str);
    int i, j;

    for (i = 0, j = length - 1; i < j; i++, j--) {
        char temp = str[i];
        str[i] = str[j];
        str[j] = temp;
    }
}

int main() {
    char str[] = "Hello, world!";
    printf("Original string: %s\n", str);

    reverseString(str);
    printf("Reversed string: %s\n", str);

    return 0;
}
```

This function does not have any comments generated, so ChatGPT uses the summarization pipeline to describe the model in the following way:

```
This code defines a function called reverseString that takes a
character array (char*) as input and reverses the string in-place. It
uses two pointers, i and j, starting from the beginning and end of the
string respectively, and swaps the characters until they meet in the
middle.

In the main function, a sample string "Hello, world!" is provided, and
the reverseString function is called to reverse it. The original and
reversed strings are then printed for verification.
```

This explanation is supposed to replace the documentation for the code, which needs to be added for good programming practice.

Finally, the tool provides us with an example of the code that tests this function:

```
Original string: Hello, world!
Reversed string: !dlrow ,olleH
```

The last part of the response is also generated as part of the prompt. This is because the test code is usually present in similar situations in the dataset. Now, if our prompt is something a bit more than just reversing a string, maybe in a different programming language, such as C#, we may get a different structure of the response. For example, if our prompt is Write a function that creates

`3000 random numbers in C# and prints them to the console`, then the response will only contain the C# code for the function, not for the test code:

```
using System;

class Program {
    static void Main(string[] args) {
        const int numNumbers = 3000;
        const int minValue = 0;
        const int maxValue = 100;

        Random random = new Random();
        for (int i = 0; i < numNumbers; i++) {
            int num = random.Next(minValue, maxValue + 1);
            Console.WriteLine(num);
        }
    }
}
```

The explanations are also generated, but not the code that tests this function. Instead, the function is wrapped as `class Program` and there is no `main()` function to test it.

Using language models in software systems

Using products such as ChatGPT is great, but they are also limited to the purpose for which they were designed. Now, we can use models like this from scratch using the Hugging Face interface. In the following code example, we can see how we can use a model dedicated to a specific task – recognizing design patterns – to complete the text – that is, writing the signature of a Singleton design pattern. This illustrates how language models (including GPT-3/4) work with text under the hood.

In the following code fragment, we're importing the model from the Hugging Face library and instantiating it. The model has been pre-trained on a set of dedicated singleton programs and constructed synthetically by adding random code from the Linux kernel source code as code of a Singleton class in C++:

```
# import the model via the huggingface library
from transformers import AutoTokenizer, AutoModelForMaskedLM


# load the tokenizer and the model for the pretrained SingBERTa
tokenizer = AutoTokenizer.from_pretrained('mstaron/SingBERTa')


# load the model
```

```
model = AutoModelForMaskedLM.from_pretrained("mstaron/SingBERTa")
```

```
# import the feature extraction pipeline
from transformers import pipeline
```

This code imports the necessary modules from the Transformers library from Hugging Face. Then, it loads the tokenizer and the model for the pre-trained SingBERTa. The tokenizer is responsible for converting text into numerical tokens, and the model is a pre-trained language model specifically designed for **masked language modeling** (**MLM**) tasks. It loads the model from the pre-trained SingBERTa. After, it imports the feature extraction pipeline from the Transformers library. The feature extraction pipeline allows us to easily extract contextualized embeddings from the model.

Overall, this code sets up the necessary components for us to use the SingBERTa model for various natural language processing tasks, such as text tokenization, MLM, and feature extraction. The following code fragment does just that – it creates the pipeline for filling in the blanks. This means that the model is prepared to predict the next word in the sentence:

```
fill_mask = pipeline(
    "fill-mask",
    model="./SingletonBERT",
    tokenizer="./SingletonBERT"
)
```

We can use this pipeline by using the `fill_mask("static Singleton:: <mask>")` command, which results in the following output:

```
[{'score': 0.9703333973884583, 'token': 74, 'token_str': 'f',
'sequence': 'static Singleton::f'},
{'score': 0.025934329256415367, 'token': 313, 'token_str': ' );',
'sequence': 'static Singleton:: );'},
{'score': 0.0003994493163190782, 'token': 279, 'token_str': '();',
'sequence': 'static Singleton::();'},
{'score': 0.00021698368072975427, 'token': 395, 'token_str': '
instance', 'sequence': 'static Singleton:: instance'},
{'score': 0.00016094298916868865, 'token': 407, 'token_str': '
getInstance', 'sequence': 'static Singleton:: getInstance'}]
```

The preceding output shows that the best prediction is the `f` token. This is correct since the training example used `f` as the name of the functions that were synthetically added to the Singleton class (`Singleton::f1()`, for example).

If we want to expand these predictions, just like the ChatGPT code generation feature, we need to loop the preceding code and generate one token at a time, thus filling in the program. There is no guarantee that the program will compile, so post-processing could essentially select only these constructs (from the list of tokens provided), which would lead to a compiling piece of code. We could even add features for testing this code, thus making our product smarter and smarter, without the need to create a larger model.

Hence, here is my last best practice for this chapter.

> **Best practice #53**
> Instead of looking for more complex models, create a smarter pipeline.

Working with a good pipeline can make a good model into a great software product. By providing the right prompt (the beginning of the text to make the prediction), we can create an output that is useful for the use case that our product fulfills.

Summary

In this chapter, we got a glimpse of what machine learning models look like from the inside, at least from the perspective of a programmer. This illustrated the major differences in how we construct machine learning-based software.

In classical models, we need to create a lot of pre-processing pipelines so that the model gets the right input. This means that we need to make sure that the data has the right properties and is in the right format; we need to work with the output to turn the predictions into something more useful.

In deep learning models, the data is pre-processed in a more streamlined way. The models can prepare the images and the text. Therefore, the software engineers' task is to focus on the product and its use case rather than monitoring concept drift, data preparation, and post-processing.

In the next chapter, we'll continue looking at examples of training machine learning models – both the classical ones and, most importantly, the deep learning ones.

References

- *Staron, M. and W. Meding. Short-term defect inflow prediction in large software project-an initial evaluation. In International Conference on Empirical Assessment in Software Engineering (EASE). 2007.*

- *Prykhodko, S. Developing the software defect prediction models using regression analysis based on normalizing transformations. In Modern problems in testing of the applied software (PTTAS-2016), Abstracts of the Research and Practice Seminar, Poltava, Ukraine. 2016.*

- *Ochodek, M., et al., Chapter 8 Recognizing Lines of Code Violating Company-Specific Coding Guidelines Using Machine Learning. In Accelerating Digital Transformation: 10 Years of Software Center. 2022, Springer. p. 211-251.*

- Ibrahim, D.R., R. Ghnemat, and A. Hudaib. *Software defect prediction using feature selection and random forest algorithm.* In *2017 International Conference on New Trends in Computing Sciences (ICTCS).* 2017. IEEE.

- Ochodek, M., M. Staron, and W. Meding, *Chapter 9 SimSAX: A Measure of Project Similarity Based on Symbolic Approximation Method and Software Defect Inflow.* In *Accelerating Digital Transformation: 10 Years of Software Center.* 2022, Springer. p. 253-283.

- Phan, V.A., *Learning Stretch-Shrink Latent Representations With Autoencoder and K-Means for Software Defect Prediction.* IEEE Access, 2022. 10: p. 117827-117835.

- Staron, M., et al., *Machine learning to support code reviews in continuous integration.* Artificial Intelligence Methods For Software Engineering, 2021: p. 141-167.

- Li, J., et al. *Software defect prediction via convolutional neural network.* In *2017 IEEE International Conference on Software Quality, Reliability and Security (QRS).* 2017. IEEE.

- Feurer, M., et al., *Efficient and robust automated machine learning.* Advances in neural information processing systems, 2015. 28.

- Feurer, M., et al., *Auto-sklearn 2.0: Hands-free automl via meta-learning.* The Journal of Machine Learning Research, 2022. 23(1): p. 11936-11996.

- Redmon, J., et al. *You only look once: Unified, real-time object detection.* In *Proceedings of the IEEE Conference on Computer Vision and Pattern Recognition.* 2016.

- Staron, M., *Automotive software architectures.* 2021: Springer.

- Gamma, E., et al., *Design patterns: elements of reusable object-oriented software.* 1995: Pearson Deutschland GmbH.

## 10

Training and Evaluating Classical Machine Learning Systems and Neural Networks

Modern machine learning frameworks are designed to be user-friendly for programmers. The popularity of the Python programming environment (and R) has shown that designing, developing, and testing machine learning models can be focused on the machine learning task and not on the programming tasks. The developers of the machine learning models can focus on developing the entire system and not on programming the internals of the algorithms. However, this bears a darker side – a lack of understanding of the internals of the models and how they are trained, evaluated, and validated.

In this chapter, I'll dive a bit deeper into the process of training and evaluation. We'll start with the basic theory behind different algorithms before learning how they are trained. We'll start with the classical machine learning models, exemplified by decision trees. Then, we'll gradually move toward deep learning, where we'll explore both dense neural networks and more advanced types of networks.

The most important part of this chapter is understanding the difference between training/evaluating algorithms and testing/validating the entire machine learning software system. I'll explain this by describing how machine learning algorithms are used as part of a production machine learning system (or what the entire machine learning system looks like).

In this chapter, we're going to cover the following main topics:

- Training and test processes
- Training classical machine learning models
- Training deep learning models
- Misleading results – problems with data leaks

Training and testing processes

Machine learning has revolutionized the way we solve complex problems by enabling computers to learn from data and make predictions or decisions without being explicitly programmed. One crucial aspect of machine learning is training models, which involves teaching algorithms to recognize patterns and relationships in data. Two fundamental methods for training machine learning models are `model.fit()` and `model.predict()`.

The `model.fit()` function lies at the heart of training a machine learning model. It is the process by which a model learns from a labeled dataset to make accurate predictions. During training, the model adjusts its internal parameters to minimize the discrepancy between its predictions and the true labels in the training data. This iterative optimization process, often referred to as "learning," allows the model to generalize its knowledge and perform well on unseen data.

In addition to the training data and labels, the `model.fit()` function also takes various hyperparameters as arguments. These hyperparameters include the number of epochs (that is, the number of times the model will iterate over the entire dataset), the batch size (the number of samples processed before updating the parameters), and the learning rate (determining the step size for parameter updates). Properly tuning these hyperparameters is crucial to ensure effective training and prevent issues such as overfitting or underfitting.

Once the training process is complete, the trained model can be used to make predictions on new, unseen data. This is where the `model.predict()` method comes into play. Given a trained model and a set of input data, the `model.predict()` function applies the learned weights and biases to generate predictions or class probabilities. The predicted outputs can then be used for various purposes, such as classification, regression, or anomaly detection, depending on the nature of the problem at hand.

We saw examples of this interface in previous chapters. Now, it is time to understand what's under the hood of this interface and how the process of training works. In the previous chapter, we looked at this process as a black box, where the process was done once the program moved past the line with `model.fit()`. This is the basics of the process, but it is not only that. The process is iterative and depends on the algorithm/model that is being trained. As every model has different parameters, the fit function can take more parameters. The additional parameters can also be added when we instantiate the model, even before the training process. *Figure 10.1* presents this process as a gray box:

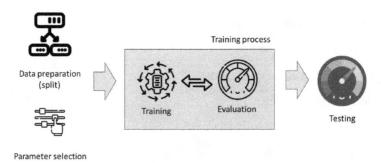

Figure 10.1 – Gray box for training a machine learning model

Before we start the training process, we split the data into training and test sets (which we discussed previously). At the same time, we select the parameters for the machine learning model that we use. These can be anything from the number of trees (for random forest) to the number of iterations and batch size for neural networks.

The training process is iterative, where a model is trained on the data, evaluated internally, and then re-trained to find a better fit for the data. In this chapter, we'll explore how this internal training works.

Finally, once the model has been trained, it is set for the testing process. In the testing process, we use pre-defined performance measures to check how well the model's learned pattern can be reproduced for new data.

Training classical machine learning models

We'll start by training a model that lets us look inside it. We'll use the CART decision tree classifier, where we can visualize the actual decision tree that is trained. We'll use the same numerical data we used in the previous chapter. First, let's read the data and create the train/test split:

```
# read the file with data using openpyxl
import pandas as pd

# we read the data from the excel file,
# which is the defect data from the ant 1.3 system
dfDataAnt13 = pd.read_excel('./chapter_6_dataset_numerical.xlsx',
                            sheet_name='ant_1_3',
                            index_col=0)

# prepare the dataset
import sklearn.model_selection

X = dfDataAnt13.drop(['Defect'], axis=1)
y = dfDataAnt13.Defect

X_train, X_test, y_train, y_test = \
        sklearn.model_selection.train_test_split(X, y, random_
state=42, train_size=0.9)
```

The preceding code reads an Excel file named `'chapter_6_dataset_numerical.xlsx'` using the `pd.read_excel()` function from pandas. The file is read into a DataFrame called `dfDataAnt13`. The `sheet_name` parameter specifies the sheet within the Excel file to read, while the `index_col` parameter sets the first column as the index of the DataFrame.

The code prepares the dataset for training a machine learning model. It assigns the independent variables (features) to the X variable by dropping the `'Defect'` column from the `dfDataAnt13` DataFrame using the `drop()` method. The dependent variable (target) is assigned to the y variable by selecting the `'Defect'` column from the `dfDataAnt13` DataFrame.

The `sklearn.model_selection.train_test_split()` function is used to split the dataset into training and testing sets. The `X` and `y` variables are split into `X_train`, `X_test`, `y_train`, and `y_test` variables. The `train_size` parameter is set to `0.9`, indicating that 90% of the data will be used for training and the remaining 10% will be used for testing. The `random_state` parameter is set to `42` to ensure reproducibility of the split.

Once the data has been prepared, we can import the decision tree library and train the model:

```
# now that we have the data prepared
# we import the decision tree classifier and train it
from sklearn.tree import DecisionTreeClassifier

# first we create an empty classifier
decisionTreeModel = DecisionTreeClassifier()

# then we train the classifier
decisionTreeModel.fit(X_train, y_train)

# and we test it for the test set
y_pred_cart = decisionTreeModel.predict(X_test)
```

The preceding code fragment imports the `DecisionTreeClassifier` class from the `sklearn.tree` module. An empty decision tree classifier object is created and assigned to the `decisionTreeModel` variable. This object will be trained on the dataset that was prepared in the previous fragment. The `fit()` method is called on the `decisionTreeModel` object to train the classifier. The `fit()` method takes the training data (`X_train`) and the corresponding target values (`y_train`) as input. The classifier will learn patterns and relationships in the training data to make predictions.

The trained decision tree classifier is used to predict the target values for the test dataset (`X_test`). The `predict()` method is called on the `decisionTreeModel` object, passing `X_test` as the input. The predicted target values are stored in the `y_pred_cart` variable. The predicted model needs to be evaluated, so let's evaluate the accuracy, precision, and recall of the model:

```
# now, let's evaluate the code
from sklearn.metrics import accuracy_score
from sklearn.metrics import recall_score
from sklearn.metrics import precision_score

print(f'Accuracy: {accuracy_score(y_test, y_pred_cart):.2f}')
print(f'Precision: {precision_score(y_test, y_pred_cart,
average="weighted"):.2f}, Recall: {recall_score(y_test, y_pred_cart,
average="weighted"):.2f}')
```

This code fragment results in the following output:

```
Accuracy: 0.83
Precision: 0.94, Recall: 0.83
```

The metrics show that the model is not that bad. It classified 83% of the data in the test set correctly. It is a bit more sensitive to the true positives (higher precision) than to true negatives (lower recall). This means that it tends to miss some of the defect-prone modules in its predictions. However, the decision tree model lets us take a look inside the model and explore the pattern that it learned from the data. The following code fragment does this:

```
from sklearn.tree import export_text

tree_rules = export_text(decisionTreeModel, feature_names=list(X_
train.columns))

print(tree_rules)
```

The preceding code fragment exports the decision tree in the form of text that we print. The `export_text()` function takes two arguments – the first one is the decision tree to visualize and the next one is the list of features. In our case, the list of features is the list of columns in the dataset.

The entire decision tree is quite complex in this case, but the first decision path looks like this:

```
|--- WMC <= 36.00
|   |--- ExportCoupling <= 1.50
|   |   |--- NOM <= 2.50
|   |   |   |--- NOM <= 1.50
|   |   |   |   |--- class: 0
|   |   |   |--- NOM >  1.50
|   |   |   |   |--- WMC <= 5.50
|   |   |   |   |   |--- class: 0
|   |   |   |   |--- WMC >  5.50
|   |   |   |   |   |--- CBO <= 4.50
|   |   |   |   |   |   |--- class: 1
|   |   |   |   |   |--- CBO >  4.50
|   |   |   |   |   |   |--- class: 0
|   |   |--- NOM >  2.50
|   |   |   |--- class: 0
```

This decision path looks very similar to a large `if-then` statement, which we could write ourselves if we knew the patterns in the data. This pattern is not simple, which means that the data is quite complex. It can be non-linear and requires complex models to capture the dependencies. It can also require a lot of effort to find the right balance between the performance of the model and its ability to generalize the data.

So, here is my best practice for working with this kind of model.

> **Best practice #54**
>
> If you want to understand your numerical data, use models that provide explainability.

In the previous chapters, I advocated for using AutoML models as they are robust and save us a lot of trouble finding the right module. However, if we want to understand our data a bit better and understand the patterns, we can start with models such as decision trees. Their insight into the data provides us with a good overview of what we can get out of the data.

As a counter-example, let's look at the data from another module from the same dataset. Let's read it and perform the split:

```python
# read the file with data using openpyxl
import pandas as pd

# we read the data from the excel file,
# which is the defect data from the ant 1.3 system
dfDataCamel12 = pd.read_excel('./chapter_6_dataset_numerical.xlsx',
                              sheet_name='camel_1_2',
                              index_col=0)

# prepare the dataset
import sklearn.model_selection

X = dfDataCamel12.drop(['Defect'], axis=1)
y = dfDataCamel12.Defect

X_train, X_test, y_train, y_test = \
        sklearn.model_selection.train_test_split(X, y, random_
state=42, train_size=0.9)
```

Now, let's train a new model for that data:

```python
# now that we have the data prepared
# we import the decision tree classifier and train it
from sklearn.tree import DecisionTreeClassifier

# first we create an empty classifier
decisionTreeModelCamel = DecisionTreeClassifier()

# then we train the classifier
decisionTreeModelCamel.fit(X_train, y_train)
```

```
# and we test it for the test set
y_pred_cart_camel = decisionTreeModel.predict(X_test)
```

So far, so good – no errors, no problems. Let's check the performance of the model:

```
# now, let's evaluate the code
from sklearn.metrics import accuracy_score
from sklearn.metrics import recall_score
from sklearn.metrics import precision_score

print(f'Accuracy: {accuracy_score(y_test, y_pred_cart_camel):.2f}')
print(f'Precision: {precision_score(y_test, y_pred_cart_camel,
average="weighted"):.2f}, Recall: {recall_score(y_test, y_pred_cart_
camel, average="weighted"):.2f}')
```

The performance, however, is not as high as it was previously:

```
Accuracy: 0.65
Precision: 0.71, Recall: 0.65
```

Now, let's print the tree:

```
from sklearn.tree import export_text

tree_rules = export_text(decisionTreeModel, feature_names=list(X_
train.columns))

print(tree_rules)
```

As we can see, the results are also quite complex:

```
|--- WMC >  36.00
|   |--- DCC <= 3.50
|   |   |--- WMC <= 64.50
|   |   |   |--- NOM <= 17.50
|   |   |   |   |--- ImportCoupling <= 7.00
|   |   |   |   |   |--- NOM <= 6.50
|   |   |   |   |   |   |--- class: 0
|   |   |   |   |   |--- NOM >  6.50
|   |   |   |   |   |   |--- CBO <= 4.50
|   |   |   |   |   |   |   |--- class: 0
|   |   |   |   |   |   |--- CBO >  4.50
|   |   |   |   |   |   |   |--- ExportCoupling <= 13.00
|   |   |   |   |   |   |   |   |--- NOM <= 16.50
|   |   |   |   |   |   |   |   |   |--- class: 1
|   |   |   |   |   |   |   |   |--- NOM >  16.50
```

```
|   |   |   |   |   |   |   |   |   |--- class: 0
|   |   |   |   |   |   |   |--- ExportCoupling >  13.00
|   |   |   |   |   |   |   |   |--- class: 0
|   |   |   |   |--- ImportCoupling >  7.00
|   |   |   |   |   |--- class: 0
|   |   |   |--- NOM >  17.50
|   |   |   |   |--- class: 1
|   |   |--- WMC >  64.50
|   |   |   |--- class: 0
```

If we look at the very first decision in this tree and the previous one, it is based on the WMC feature. **WMC** means **weighted method per class** and is one of the classical software metrics that was introduced in the 1990s by Chidamber and Kamerer. The metric captures both the complexity and the size of the class (in a way) and it is quite logical that large classes are more defect-prone – simply because there is more chance to make a mistake if there is more source code. In the case of this model, this is a bit more complicated as the model recognizes that the classes with WMC over 36 are more prone to errors than others, apart from classes that are over 64.5, which are less prone to errors. The latter is also a known phenomenon that large classes are also more difficult to test and therefore can contain undiscovered defects.

Here is my next best practice, which is about the explainability of models.

> **Best practice #55**
> The best models are those that capture the empirical phenomena in the data.

Although machine learning models can capture any kind of dependencies, the best models are the ones that can capture logical, empirical observations. In the previous examples, the model could capture the software engineering empirical observations related to the size of the classes and their defect-proneness. Having a model that captures empirical relations leads to better products and explainable AI.

Understanding the training process

From the software engineer's perspective, the training process is rather simple – we fit the model, validate it, and use it. We check how good the model is in terms of the performance metrics. If the model is good enough, and we can explain it, then we develop the entire product around it, or we use it in a larger software product.

When the model does not learn anything useful, we need to understand why this is the case and whether there could be another model that can. We can use the visualization techniques we learned about in *Chapter 6* to explore the data and clear it from noise using the techniques from *Chapter 4*.

Now, let's explore the process of how the decision tree model learns from the data. The DecisionTree classifier learns from the provided data by recursively partitioning the feature space based on the values of the features in the training dataset. It constructs a binary tree where each internal node represents a feature and a decision rule based on a threshold value, and each leaf node represents a predicted class or outcome.

The training is done in steps:

1. **Selecting the best feature**: The classifier evaluates different features and determines the one that best separates the data into different classes. This is typically done using a measure of impurity or information gain, such as Gini impurity or entropy.

2. **Splitting the dataset**: Once the best feature has been selected, the classifier splits the dataset into two or more subsets based on the values of that feature. Each subset represents a different branch or path in the decision tree.

3. **Repeating the process recursively**: The preceding steps are repeated for each subset or branch of the decision tree, treating them as separate datasets. The process continues until a stopping condition is met, such as reaching a maximum depth, a minimum number of samples at a node, or other predefined criteria.

4. **Assigning class labels**: At the leaf nodes of the decision tree, the classifier assigns class labels based on the majority class of the samples in that region. This means that when making predictions, the classifier assigns the most frequent class in the leaf node to the unseen samples that fall into that region.

During the learning process, the `DecisionTree` classifier aims to find the best splits that maximize the separation of classes and minimize the impurity within each resulting subset. By recursively partitioning the feature space based on the provided training data, the classifier learns decision rules that allow it to make predictions for unseen data.

It's important to note that decision trees are prone to overfitting, meaning they can memorize the training data too well and not generalize well to new data. Techniques such as pruning, limiting the maximum depth, or using ensemble methods such as random forest can help mitigate overfitting and improve the performance of decision tree models.

We used the random forest classifier previously in this book, so we won't dive into the details here. Although random forests are better at generalizing data, they are opaque compared to decision trees. We cannot explore what the model learned – we can only explore which features contribute the most to the verdict.

Random forest and opaque models

Let's train the random forest classifier based on the same data as in the counter-example and check whether the model performs better and whether the model uses similar features as the `DecisionTree` classifier in the original counter-example.

Let's instantiate, train, and validate the model on the same data using the following fragment of code:

```
from sklearn.ensemble import RandomForestClassifier

randomForestModel = RandomForestClassifier()
randomForestModel.fit(X_train, y_train)
y_pred_rf = randomForestModel.predict(X_test)
```

After evaluating the model, we obtain the following performance metrics:

```
Accuracy: 0.62
Precision: 0.63, Recall: 0.62
```

Admittedly, these metrics are different than the metrics in the decision trees, but the overall performance is not that much different. The difference in accuracy of 0.03 is negligible. First, we can extract the important features, reusing the same techniques that were presented in *Chapter 5*:

```
# now, let's check which of the features are the most important ones
# first we create a dataframe from this list
# then we sort it descending
# and then filter the ones that are not imporatnt
dfImportantFeatures = pd.DataFrame(randomForestModel.feature_
importances_, index=X.columns, columns=['importance'])

# sorting values according to their importance
dfImportantFeatures.sort_values(by=['importance'],
                                ascending=False,
                                inplace=True)

# choosing only the ones that are important, skipping
# the features which have importance of 0
dfOnlyImportant =
dfImportantFeatures[dfImportantFeatures['importance'] != 0]

# print the results
print(f'All features: {dfImportantFeatures.shape[0]}, but only
{dfOnlyImportant.shape[0]} are used in predictions. ')
```

We can visualize the set of features used in the decision by executing the following code:

```
# we use matplotlib and seaborn to make the plot
import matplotlib.pyplot as plt
import seaborn as sns

# Define size of bar plot
# We make the x axis quite much larger than the y-axis since
```

```
# there is a lot of features to visualize
plt.figure(figsize=(40,10))

# plot Searborn bar chart
# we just use the blue color
sns.barplot(y=dfOnlyImportant['importance'],
            x=dfOnlyImportant.index,
            color='steelblue')

# we make the x-labels rotated so that we can fit
# all the features
plt.xticks(rotation=90)

sns.set(font_scale=6)

# add chart labels
plt.title('Importance of features, in descending order')
plt.xlabel('Feature importance')
plt.ylabel('Feature names')
```

This code helps us understand the importance chart shown in *Figure 10.2*. Here, again, the WMC is the most important feature. This means that a lot of trees in the forest are using this metric to make decisions. However, we do not know the algorithm since the forest is an ensemble classifier – it uses voting for the decisions – meaning that always more than one tree is used when making the final call/prediction:

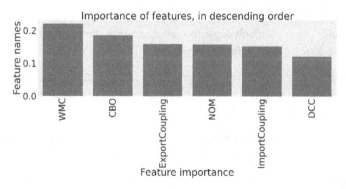

Figure 10.2 – Feature importance chart for the random forest classifier.

Please note that the model is more complex than just a linear combination of these features. This chart illustrates something that is not a best practice, but a best experience. So, I will use it as a best practice to illustrate the importance of it.

> **Best practice #56**
>
> Simple, but explainable, models can often capture data in a good way.

What I've learned throughout my experiments with different types of data is that if there is a pattern, a simple model will capture it. If there is no pattern, or if the data has a lot of exceptions from rules, then even the most complex models will have problems in finding the patterns. Therefore, if you cannot explain your results, do not use them in your product, as there is a chance that these results will make the products quite useless.

However, there is a light at the end of the tunnel here. Some models can capture very complex patterns, but they are opaque – neural networks.

Training deep learning models

Training a dense neural network involves various steps. First, we prepare the data. This typically involves tasks such as feature scaling, handling missing values, encoding categorical variables, and splitting the data into training and validation sets.

Then, we define the architecture of the dense neural network. This includes specifying the number of layers, the number of neurons in each layer, the activation functions to be used, and any regularization techniques, such as dropout or batch normalization.

Once the model has been defined, we need to initialize it. We create an instance of the neural network model based on the defined architecture. This involves creating an instance of the neural network class or using a predefined model architecture available in a deep learning library. We also need to define a loss function that quantifies the error between the predicted output of the model and the actual target values. The choice of loss function depends on the nature of the problem, such as classification (cross-entropy) or regression (mean squared error).

In addition to the loss function, we need an optimizer. The optimizer algorithm will update the weights of the neural network during training. Common optimizers include **stochastic gradient descent** (**SGD**), Adam, and RMSprop.

Then, we can train the model. Here, we iterate over the training data for multiple epochs (passes through the entire dataset). In each epoch, perform the following steps:

1. **Forward pass**: We feed a batch of input data into the model and compute the predicted output.

2. **Compute loss**: We compare the predicted output with the actual target values using the defined loss function to calculate the loss.

3. **Backward pass**: We propagate the loss backward through the network to compute the gradients of the weights concerning the loss using backpropagation.

4. **Update the weights**: We use the optimizer to update the weights of the neural network based on the computed gradients, adjusting the network parameters to minimize the loss.

We repeat these steps for each batch in the training data until all batches have been processed.

Finally, we need to perform the validation process, just like in the previous models. Here, we compute a validation metric (such as accuracy or mean squared error) to assess how well the model is generalizing to unseen data. This helps us monitor the model's progress and detect overfitting.

Once the model has been trained and validated, we can evaluate its performance on a separate test dataset that was not used during training or validation. Here, we calculate relevant evaluation metrics to assess the model's accuracy, precision, recall, or other desired metrics.

So, let's do this for our dataset. First, we must define the architecture of the model using the following code:

```python
import torch
import torch.nn as nn
import torch.optim as optim

# Define the neural network architecture
class NeuralNetwork(nn.Module):
    def __init__(self, input_size, hidden_size, num_classes):
        super(NeuralNetwork, self).__init__()
        self.fc1 = nn.Linear(input_size, hidden_size)
        self.relu = nn.ReLU()
        self.fc2 = nn.Linear(hidden_size, num_classes)

    def forward(self, x):
        out = self.fc1(x)
        out = self.relu(out)
        out = self.fc2(out)
        return out
```

Here, we define a class called `NeuralNetwork`, which is a subclass of `nn.Module`. This class represents our neural network model. It has two fully connected layers (`fc1` and `fc2`) with a ReLU activation function in between. The network looks something like the one shown in *Figure 10.3*:

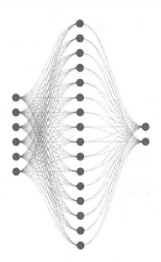

Figure 10.3 – Neural network used for predicting defects.

This visualization was created using `http://alexlenail.me/NN-SVG/index.html`. The number of neurons in the hidden layer is 64, but in this figure, only 16 are shown to make it more readable. The network starts with 6 input neurons, then 64 neurons in the hidden layer (middle), and two neurons for the decision classes at the end.

Now, we can define the hyperparameters for training the network and instantiate it:

```
# Define the hyperparameters
input_size = X_train.shape[1]  # Number of input features
hidden_size = 64               # Number of neurons in the hidden layer
num_classes = 2                # Number of output classes

# Create an instance of the neural network
model = NeuralNetwork(input_size, hidden_size, num_classes)

# Define the loss function and optimizer
criterion = nn.CrossEntropyLoss()
optimizer = optim.Adam(model.parameters(), lr=0.001)

# Convert the data to PyTorch tensors
X_train_tensor = torch.Tensor(X_train.values)
y_train_tensor = torch.LongTensor(y_train.values)
X_test_tensor = torch.Tensor(X_test.values)

# Training the neural network
num_epochs = 10000
batch_size = 32
```

Here, we create an instance of the `NeuralNetwork` class called `model` with the specified input size, hidden size, and number of output classes, as we defined in the first code fragment. We define the loss function (cross-entropy loss) and the optimizer (Adam optimizer) to train the model. The data is then converted into PyTorch tensors using `torch.Tensor()` and `torch.LongTensor()`. Finally, we say that we want to train the model in 10,000 epochs (iterations) with 32 elements (data points) in each iteration:

```
for epoch in range(num_epochs):
    for I in range(0, len(X_train_tensor), batch_size):
        batch_X = X_train_tensor[i:i+batch_size]
        batch_y = y_train_tensor[i:i+batch_size]

        # Forward pass
        outputs = model(batch_X)
        loss = criterion(outputs, batch_y)

        # Backward and optimize
        optimizer.zero_grad()
        loss.backward()
        optimizer.step()

    # Print the loss at the end of each epoch
    if (epoch % 100 == 0):
      print(""Epoch {epoch+1}/{num_epochs}, Loss: {loss.item():.3f"")
```

Now, we can get the predictions for the test data and obtain the performance metrics:

```
with torch.no_grad():
    model.eval()  # Set the model to evaluation mode
    X_test_tensor = torch.Tensor(X_test.values)
    outputs = model(X_test_tensor)
    _, predicted = torch.max(outputs.data, 1)
    y_pred_nn = predicted.numpy()

# now, let's evaluate the code
from sklearn.metrics import accuracy_score
from sklearn.metrics import recall_score
from sklearn.metrics import precision_score

print(f'Accuracy: {accuracy_score(y_test, y_pred_nn):.2f}')
print(f'Precision: {precision_score(y_test, y_pred_nn,
average="weighted"):.2f}, Recall: {recall_score(y_test, y_pred_nn,
average="weighted"):.2f}')
```

The performance metrics are as follows:

```
Accuracy: 0.73
Precision: 0.79, Recall: 0.73
```

So, this is a bit better than the previous models, but it's not great. The patterns are just not there. We could make the network larger by increasing the number of hidden layers, but this does not make the predictions better.

Misleading results – data leaking

In the training process, we use one set of data and in the test set, we use another set. The best training process is when these two datasets are separate. If they are not, we get into something that is called a data leak problem. This problem is when we have the same data points in both the train and test sets. Let's illustrate this with an example.

First, we need to create a new split, where we have some data points in both sets. We can do that by using the split function and setting 20% of the data points to the test set. This means that at least 10% of the data points are in both sets:

```
X_trainL, X_testL, y_trainL, y_testL = \
        sklearn.model_selection.train_test_split(X, y, random_
state=42, train_size=0.8)
```

Now, we can use the same code to make predictions on this data and then calculate the performance metrics:

```
# now, let's evaluate the model on this new data
with torch.no_grad():
    model.eval()  # Set the model to evaluation mode
    X_test_tensor = torch.Tensor(X_testL.values)
    outputs = model(X_test_tensor)
    _, predicted = torch.max(outputs.data, 1)
    y_pred_nn = predicted.numpy()

print(f'Accuracy: {accuracy_score(y_testL, y_pred_nn):.2f}')
print(f'Precision: {precision_score(y_testL, y_pred_nn,
average="weighted"):.2f}, Recall: {recall_score(y_testL, y_pred_nn,
average="weighted"):.2f}')
```

The results are as follows:

```
Accuracy: 0.85
Precision: 0.86, Recall: 0.85
```

The results are better than before. However, they are only better because 10% of the data points were used in both the training and the test sets. This means that the model performs much worse than the metrics suggest. Hence, we've come to my next best practice.

> **Best practice #56**
> Always make sure that the data points in both the train and test sets are separate.

Although we made this mistake on purpose here, it is quite easy to make it in practice. Please note the `random_state=42` parameter in the split function. Setting it explicitly ensures that the split is repeatable. However, if we do not do this, we can end up with different splits every time we make them and thus we can end up with the data leak problem.

The data leak problem is even more difficult to discover when we're working with images or text. Just the fact that an image comes from two different files does not guarantee that it is different. For example, images taken one after another during highway driving will be different but will not be too different, and if they end up in test and train sets, we get a whole new dimension of the data leak problem.

Summary

In this chapter, we discussed various topics related to machine learning and neural networks. We explained how to read data from an Excel file using the pandas library and prepare the dataset for training a machine learning model. We explored the use of decision tree classifiers and demonstrated how to train a decision tree model using scikit-learn. We also showed how to make predictions using the trained model.

Then, we discussed how to switch from a decision tree classifier to a random forest classifier, which is an ensemble of decision trees. We explained the necessary code modifications and provided an example. Next, we shifted our focus to using a dense neural network in PyTorch. We described the process of creating the neural network architecture, training the model, and making predictions using the trained model.

Lastly, we explained the steps involved in training a dense neural network, including data preparation, model architecture, initializing the model, defining a loss function and optimizer, the training loop, validation, hyperparameter tuning, and evaluation.

Overall, we covered a range of topics related to machine learning algorithms, including decision trees, random forests, and dense neural networks, along with their respective training processes.

In the next chapter we explore how to train more advanced machine learning models - for example AutoEncoders.

References

- *Chidamber, S.R. and C.F. Kemerer, A metrics suite for object oriented design. IEEE Transactions on Software Engineering, 1994. 20(6): p. 476–493.*

11

Training and Evaluation of Advanced ML Algorithms – GPT and Autoencoders

Classical **machine learning** (**ML**) and **neural networks** (**NNs**) are very good for classical problems – prediction, classification, and recognition. As we learned in the previous chapter, training them requires a moderate amount of data, and we train them for specific tasks. However, breakthroughs in ML and **artificial intelligence** (**AI**) in the late 2010s and the beginning of 2020s were about completely different types of models – **deep learning** (**DL**), **Generative Pre-Trained Transformers** (**GPTs**), and **generative AI** (**GenAI**).

GenAI models provide two advantages – they can generate new data and they can provide us with an internal representation of the data that captures the context of the data and, to some extent, its semantics. In the previous chapters, we saw how we can use existing models for inference and generating simple pieces of text.

In this chapter, we explore how GenAI models work based on GPT and Bidirectional Encoder Representations from Transformers (BERT) models. These models are designed to generate new data based on the patterns that they were trained on. We also look at the concept of autoencoders (AEs), where we train an AE to generate new images based on previously trained data.

In this chapter, we're going to cover the following main topics:

- From classical ML models to GenAI
- The theory behind GenAI models – AEs and transformers
- Training and evaluation of a **Robustly Optimized BERT Approach** (**RoBERTa**) model
- Training and evaluation of an AE
- Developing safety cages to prevent models from breaking the entire system

From classical ML to GenAI

Classical AI, also known as symbolic AI or rule-based AI, emerged as one of the earliest schools of thought in the field. It is rooted in the concept of explicitly encoding knowledge and using logical rules to manipulate symbols and derive intelligent behavior. Classical AI systems are designed to follow predefined rules and algorithms, enabling them to solve well-defined problems with precision and determinism. We delve into the underlying principles of classical AI, exploring its reliance on rule-based systems, expert systems, and logical reasoning.

In contrast, GenAI represents a paradigm shift in AI development, capitalizing on the power of ML and NNs to create intelligent systems that can generate new content, recognize patterns, and make informed decisions. Rather than relying on explicit rules and handcrafted knowledge, GenAI leverages data-driven approaches to learn from vast amounts of information and infer patterns and relationships. We examine the core concepts of GenAI, including DL, NNs, and probabilistic models, to unravel its ability to create original content and foster creative problem-solving.

One of the examples of a GenAI model is the GPT-3 model. GPT-3 is a state-of-the-art language model developed by OpenAI. It is based on the transformer architecture. GPT-3 is trained using a technique called **unsupervised learning (UL)**, which enables it to generate coherent and contextually relevant text.

The theory behind advanced models – AEs and transformers

One of the large limitations of classical ML models is the access to annotated data. Large NNs contain millions (if not billions) of parameters, which means that they require equally many labeled data points to be trained correctly. Data labeling, also known as annotation, is the most expensive activity in ML, and therefore it is the labeling process that becomes the de facto limit of ML models. In the early 2010s, the solution to that problem was to use crowdsourcing.

Crowdsourcing, which is a process of collective data collection (among other things), means that we use users of our services to label the data. A CAPTCHA is one of the most prominent examples. A CAPTCHA is used when we need to recognize images in order to log in to a service. When we introduce new images, every time a user needs to recognize these images, we can label a lot of data in a relatively short time.

There is, nevertheless, an inherent problem with that process. Well, there are a few problems, but the most prominent one is that this process works mostly with images or similar kinds of data. It is also a relatively limited process – we can only ask users to recognize an image, but not add a semantic map and not draw a bounding box over an image. We cannot ask users to assess the similarity of images or any other, bit more advanced, task.

Here enter more advanced methods – GenAI and networks such as **generative adversarial networks (GANs)**. These networks are designed to generate data and learn which data is like the original data.

These networks are very powerful and have been used in such applications as the generation of images; for example, in the so-called "deep fakes."

AEs

One of the main components of such a model is the AE, which is designed to learn a compressed representation (encoding) of the input data and then reconstruct the original data (decoding) from this compressed representation.

The architecture of an AE (*Figure 11.1*) consists of two main components: an encoder and a decoder. The encoder takes the input data and maps it to a lower-dimensional latent space representation, often referred to as the encoding/embedding or latent representation. The decoder takes this encoded representation and reconstructs the original input data:

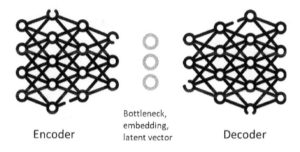

Encoder   Bottleneck, embedding, latent vector   Decoder

Figure 11.1 – High-level architecture of AEs

The objective of an AE is to minimize reconstruction errors, which is the difference between the input data and the output of the decoder. By doing so, the AE learns to capture the most important features of the input data in the latent representation, effectively compressing the information. The most interesting part is the latent space or encoding. This part allows this model to learn the representation of a complex data point (for example, an image) in a small vector of just a few numbers. The latent representation learned by the AE can be considered a compressed representation or a low-dimensional embedding of the input data. This compressed representation can be used for various purposes, such as data visualization, dimensionality reduction, anomaly detection, or as a starting point for other downstream tasks.

The encoder part calculates the latent vector, and the decoder part can expand it to an image. There are different types of AEs; the most interesting one is the **variational AE** (**VAE**), which encodes the parameters of a function that can generate new data rather than the representation of the data itself. In this way, it can create new data based on the distribution. In fact, it can even create completely new types of data by combining different functions.

Transformers

In **natural language processing** (**NLP**) tasks, we usually employ a bit different type of GenAI – transformers. Transformers revolutionized the field of machine translation but have been applied to many other tasks, including language understanding and text generation.

At its core, a transformer employs a self-attention mechanism that allows the model to weigh the importance of different words or tokens in a sequence when processing them. This attention mechanism enables the model to capture long-range dependencies and contextual relationships between words more effectively than traditional **recurrent NNs** (**RNNs**) or **convolutional NNs** (**CNNs**).

Transformers consist of an encoder-decoder structure. The encoder processes the input sequence, such as a sentence, and the decoder generates an output sequence, often based on the input and a target sequence. Two elements are unique for transformers:

- **Multi-head self-attention (MHSA)**: A mechanism that allows the model to attend to different positions in the input sequence simultaneously, capturing different types of dependencies. This is an extension to the RNN architecture, which was able to connect neurons in the same layer, thus capturing temporal dependencies.

- **Positional encoding**: To incorporate positional information into the model, positional encoding vectors are added to the input embeddings. These positional encodings are based on the tokens and their relative position to one another. This mechanism allows us to capture the context of a specific token and therefore to capture the basic contextual semantics of the text.

Figure 11.2 presents the high-level architecture of transformers:

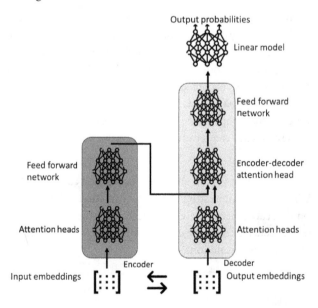

Figure 11.2 – High-level architecture of transformers

In this architecture, self-attention is a key mechanism that allows the model to weigh the importance of different words or tokens in a sequence when processing them. The self-attention mechanism is applied independently to each word in the input sequence, and it helps capture contextual relationships and dependencies between words. The term *head* refers to an independent attention mechanism that operates in parallel. Multiple self-attention heads can be used in the transformer model to capture different types of relationships (although we do not know what these relationships are).

Each self-attention head operates by computing attention scores between query representations and key representations. These attention scores indicate the importance or relevance of each word in the sequence with respect to the others. Attention scores are obtained by taking the dot product between the query and key representations, followed by applying a softmax function to normalize the scores.

The attention scores are then used to weigh the value representations. The weighted values are summed together to obtain the output representation for each word in the sequence.

The feed-forward networks in transformers serve two main purposes: feature extraction and position-wise representation. The feature extraction extracts higher-level features from the self-attention outputs, in a way that is very similar to the word-embeddings extraction that we learned previously. By applying non-linear transformations, the model can capture complex patterns and dependencies in the input sequence. The position-wise representation ensures that the model can learn different transformations for each position. It allows the model to learn complex representations of the sentences and therefore capture the more complex context of each word and sentence.

The transformer architecture is the basis for modern models such as GPT-3, which is a pre-trained generative transformer; that is, a transformer that has been pre-trained on a large mass of text. However, it is based on models such as BERT and its relatives.

Training and evaluation of a RoBERTa model

In general, the training process for GPT-3 involves exposing the model to a massive amount of text data from diverse sources, such as books, articles, websites, and more. By analyzing the patterns, relationships, and language structures within this data, the model learns to predict the likelihood of a word or phrase appearing based on the surrounding context. This learning objective is achieved through a process known as **masked language modeling** (**MLM**), where certain words are randomly masked in the input, and the model is tasked with predicting the correct word based on the context.

In this chapter, we train the RoBERTa model, which is a variation of the now-classical BERT model. Instead of using generic sources such as books and *Wikipedia* articles, we use programs. To make our training task a bit more specific, let us train a model that is capable of "understanding" code from a networking domain – WolfSSL, which is an open source implementation of the SSL protocol, used in many embedded software devices.

Once the training is complete, BERT models are capable of generating text by leveraging their learned knowledge and the context provided in a given prompt. When a user provides a prompt or a partial sentence, the model processes the input and generates a response by probabilistically predicting the most likely next word based on the context it has learned from the training data.

When it comes to GPT-3 (and similar) models, it is an extension of the BERT model. The generation process in GPT-3 involves multiple layers of attention mechanisms within the transformer architecture. These attention mechanisms allow the model to focus on relevant parts of the input text and make connections between different words and phrases, ensuring coherence and contextuality in the generated output. The model generates text by sampling or selecting the most probable next word at each step, taking into account previously generated words.

So, let us start our training process by preparing the data for training. First, we read the dataset:

```
from tokenizers import ByteLevelBPETokenizer

paths = ['source_code_wolf_ssl.txt']

print(f'Found {len(paths)} files')
print(f'First file: {paths[0]}')
```

This provides us with the raw training set. In this set, the text file contains all the source code from the WolfSSL protocol in one file. We do not have to prepare it like this, but it certainly makes the process easier as we only deal with one source file. Now, we can train the tokenizer, very similar to what we saw in the previous chapters:

```
# Initialize a tokenizer
tokenizer = ByteLevelBPETokenizer()

print('Training tokenizer...')

# Customize training
# we use a large vocabulary size, but we could also do with ca. 10_000
tokenizer.train(files=paths,
                vocab_size=52_000,
                min_frequency=2,
                special_tokens=["<s>","<pad>","</
s>","<unk>","<mask>",])
```

The first line initializes an instance of the ByteLevelBPETokenizer tokenizer class. This tokenizer is based on a byte-level version of the **Byte-Pair Encoding** (**BPE**) algorithm, which is a popular subword tokenization method. We discussed it in the previous chapters.

The next line prints a message indicating that the tokenizer training process is starting.

The `tokenizer.train()` function is called to train the tokenizer. The training process takes a few parameters:

- `files=paths`: This parameter specifies the input files or paths containing the text data to train the tokenizer. It expects a list of file paths.

- `vocab_size=52_000`: This parameter sets the size of the vocabulary; that is, the number of unique tokens the tokenizer will generate. In this case, the tokenizer will create a vocabulary of 52,000 tokens.

- `min_frequency=2`: This parameter specifies the minimum frequency a token must have in the training data to be included in the vocabulary. Tokens that occur less frequently than this threshold will be treated as **out-of-vocabulary** (**OOV**) tokens.

- `special_tokens=["<s>","<pad>","</s>","<unk>","<mask>"]`: This parameter defines a list of special tokens that will be added to the vocabulary. Special tokens are commonly used to represent specific meanings or special purposes. In this case, the special tokens are <s>, <pad>, </s>, <unk>, and <mask>. These tokens are often used in tasks such as machine translation, text generation, or language modeling.

Once the training process is completed, the tokenizer will have learned the vocabulary and will be able to encode and decode text using the trained subword units. We can now save the tokenizer using this piece of code:

```
import os

# we give this model a catchy name - wolfBERTa
# because it is a RoBERTa model trained on the WolfSSL source code
token_dir = './wolfBERTa'

if not os.path.exists(token_dir):
  os.makedirs(token_dir)

tokenizer.save_model('wolfBERTa')
```

We also test this tokenizer using the following line: `tokenizer.encode("int main(int argc, void **argv)").tokens`.

Now, let us make sure that the tokenizer is comparable with our model in the next step. To do that, we need to make sure that the output of the tokenizer never exceeds the number of tokens that the model can accept:

```
from tokenizers.processors import BertProcessing

# let's make sure that the tokenizer does not provide more tokens than
we expect
```

```
# we expect 512 tokens, because we will use the BERT model
tokenizer._tokenizer.post_processor = BertProcessing(
    ("</s>", tokenizer.token_to_id("</s>")),
    ("<s>", tokenizer.token_to_id("<s>")),
)
tokenizer.enable_truncation(max_length=512)
```

Now, we can move over to preparing the model. We do this by importing the predefined class from the HuggingFace hub:

```
import the RoBERTa configuration
from transformers import RobertaConfig

# initialize the configuration
# please note that the vocab size is the same as the one in the
tokenizer.
# if it is not, we could get exceptions that the model and the
tokenizer are not compatible
config = RobertaConfig(
    vocab_size=52_000,
    max_position_embeddings=514,
    num_attention_heads=12,
    num_hidden_layers=6,
    type_vocab_size=1,
)
```

The first line, `from transformers import RobertaConfig`, imports the `RobertaConfig` class from the `transformers` library. The `RobertaConfig` class is used to configure the RoBERTa model. Next, the code initializes the configuration of the RoBERTa model. The parameters passed to the `RobertaConfig` constructor are as follows:

- `vocab_size=52_000`: This parameter sets the size of the vocabulary used by the RoBERTa model. It should match the vocabulary size used during the tokenizer training. In this case, the tokenizer and the model both have a vocabulary size of 52,000, ensuring they are compatible.

- `max_position_embeddings=514`: This parameter sets the maximum sequence length that the RoBERTa model can handle. It defines the maximum number of tokens in a sequence that the model can process. Longer sequences may need to be truncated or split into smaller segments. Please note that the input is 514, not 512 as the output of the tokenizer. This is caused by the fact that we leave the place from the starting and ending tokens.

- `num_attention_heads=12`: This parameter sets the number of attention heads in the **multi-head attention** (**MHA**) mechanism of the RoBERTa model. Attention heads allow the model to focus on different parts of the input sequence simultaneously.

- `num_hidden_layers=6`: This parameter sets the number of hidden layers in the RoBERTa model. These layers contain the learnable parameters of the model and are responsible for processing the input data.

- `type_vocab_size=1`: This parameter sets the size of the token type vocabulary. In models such as RoBERTa, which do not use the token type (also called a segment) embeddings, this value is typically set to 1.

The configuration object config stores all these settings and will be used later when initializing the actual RoBERTa model. Having the same configuration parameters as the tokenizer ensures that the model and tokenizer are compatible and can be used together to process text data properly.

It is worth noting that this model is rather small, compared to the 175 billion parameters of GPT-3. It has (only) 85 million parameters. However, it can be trained on a laptop with a moderately powerful GPU (any NVIDIA GPU with 6 GB of VRAM will do). The model is, nevertheless, much larger than the original BERT model from 2017, which had only six attention heads and a handful of millions of parameters.

Once the model is created, we need to initiate it:

```
# Initializing a Model From Scratch
from transformers import RobertaForMaskedLM

# initialize the model
model = RobertaForMaskedLM(config=config)

# let's print the number of parameters in the model
print(model.num_parameters())

# let's print the model
print(model)
```

The last two lines print out the number of parameters in the model (a bit over 85 million) and then the model itself. The output of that model is quite large, so we do not present it here.

Now that the model is ready, we need to go back to the dataset and prepare it for training. The simplest way is to reuse the previously trained tokenizer by reading it back from the folder, but with the changed class of that tokenizer so that it fits the model:

```
from transformers import RobertaTokenizer

# initialize the tokenizer from the file
tokenizer = RobertaTokenizer.from_pretrained("./wolfBERTa", max_
length=512)
```

Once this is done, we can read the dataset:

```
from datasets import load_dataset

new_dataset = load_dataset("text", data_files='./source_code_wolf_ssl.
txt')
```

The previous code fragment reads the same dataset that we used to train the tokenizer. Now, we will use the tokenizer to transform the dataset into a set of tokens:

```
tokenized_dataset = new_dataset.map(lambda x: tokenizer(x["text"]),
num_proc=8)
```

This takes a moment, but it gives us a moment to also reflect on the fact that this code takes advantage of the so-called map-reduce algorithm, which became a golden standard for processing large files at the beginning of the 2010s when the concept of big data was very popular. It is the map () function that utilizes that algorithm.

Now, we need to prepare the dataset for training by creating so-called masked input. Masked input is a set of sentences where words are replaced by the mask token (<mask> in our case). It can look something like the example in *Figure 11.3*:

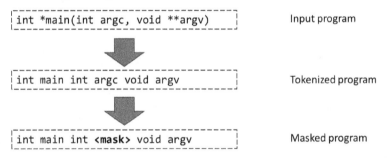

Figure 11.3 – Masked input for MLMs

It's easy to guess that the <mask> token can appear at any place and that it should appear several times in similar places in order for the model to actually learn the masked token's context. It would be very cumbersome to do it manually, and therefore, the HuggingFace library has a dedicated class for it – DataCollatorForLanguageModeling. The following code demonstrates how to instantiate that class and how to use its parameters:

```
from transformers import DataCollatorForLanguageModeling

data_collator = DataCollatorForLanguageModeling(
    tokenizer=tokenizer, mlm=True, mlm_probability=0.15
)
```

The `from transformers import DataCollatorForLanguageModeling` line imports the `DataCollatorForLanguageModeling` class, which is used for preparing data for language modeling tasks. The code initializes a `DataCollatorForLanguageModeling` object named `data_collator`. This object takes several parameters:

- `tokenizer=tokenizer`: This parameter specifies the tokenizer to be used for encoding and decoding the text data. It expects an instance of a `tokenizer` object. In this case, it appears that the `tokenizer` object has been previously defined and assigned to the `tokenizer` variable.

- `mlm=True`: This parameter indicates that the language modeling task is an MLM task.

- `mlm_probability=0.15`: This parameter sets the probability of masking a token in the input text. Each token has a 15% chance of being masked during data preparation.

The `data_collator` object is now ready to be used for preparing data for language modeling tasks. It takes care of tasks such as tokenization and masking of the input data to be compatible with the RoBERTa model. Now, we can instantiate another helper class – `Trainer` – which manages the training process of the MLM model:

```
from transformers import Trainer, TrainingArguments

training_args = TrainingArguments(
    output_dir="./wolfBERTa",
    overwrite_output_dir=True,
    num_train_epochs=10,
    per_device_train_batch_size=32,
    save_steps=10_000,
    save_total_limit=2,
)

trainer = Trainer(
    model=model,
    args=training_args,
    data_collator=data_collator,
    train_dataset=tokenized_dataset['train'],
)
```

The `from transformers import Trainer, TrainingArguments` line imports the `Trainer` class and the `TrainingArguments` class from the `transformers` library. Then it initializes a `TrainingArguments` object, `training_args`. This object takes several parameters to configure the training process:

- `output_dir="./wolfBERTa"`: This parameter specifies the directory where the trained model and other training artifacts will be saved.

- `overwrite_output_dir=True`: This parameter determines whether to overwrite `output_dir` if it already exists. If set to `True`, it will overwrite the directory.

- `num_train_epochs=10`: This parameter sets the number of training epochs; that is, the number of times the training data will be iterated during training. In our example, it is enough with a few epochs only, such as 10. It takes a lot of time to train these models, so that's why we go with a small number of epochs.

- `per_device_train_batch_size=32`: This parameter sets the batch size per GPU for training. It determines how many training examples are processed together in parallel during each training step. If you do not have a lot of VRAM in your GPU, decrease this number.

- `save_steps=10_000`: This parameter specifies the number of training steps before saving a checkpoint of the model.

- `save_total_limit=2`: This parameter limits the total number of saved checkpoints. If the limit is exceeded, older checkpoints will be deleted.

After initializing the trainer arguments, the code initializes a `Trainer` object with the following arguments:

- `model=model`: This parameter specifies the model to be trained. In this case, the pre-initialized RoBERTa model from our previous steps is assigned to the model variable.

- `args=training_args`: This parameter specifies the training arguments, which we prepared in our previous steps.

- `data_collator=data_collator`: This parameter specifies the data collator to be used during training. This object was prepared previously in our code.

- `train_dataset=tokenized_dataset['train']`: This parameter specifies the training dataset. It appears that a tokenized dataset has been prepared and stored in a dictionary called `tokenized_dataset`, and the training portion of that dataset is assigned to `train_dataset`. In our case, since we did not define the train-test split, it take the entire dataset.

The `Trainer` object is now ready to be used for training the RoBERTa model using the specified training arguments, data collator, and training dataset. We do this by simply writing `trainer.train()`.

Once the model finishes training, we can save it using the following command: `trainer.save_model("./wolfBERTa")`. After that, we can use the model just as we learned in *Chapter 10*.

It takes a while to train the model; on a consumer-grade GPU such as NVIDIA 4090, it can take about one day for 10 epochs, but if we want to use a larger dataset or more epochs, it can take much longer. I do not advise executing this code on a computer without a GPU as it takes ca. 5-10 times longer than on a GPU. Hence my next best practice.

> **Best practice #57**
>
> Use NVIDIA **Compute Unified Device Architecture** (**CUDA**; accelerated computing) for training advanced models such as BERT, GPT-3, and AEs.

For classical ML, and even for simple NNs, a modern CPU is more than enough. The number of calculations is large, but not extreme. However, when it comes to training BERT models, AEs, and similar, we need acceleration for handling tensors (vectors) and making calculations on entire vectors at once. CUDA is NVIDIA's acceleration framework. It allows developers to utilize the power of NVIDIA GPUs to accelerate computational tasks, including training DL models. It provides a few benefits:

- **GPU parallelism**, designed to handle many parallel computations simultaneously. DL models, especially large models such as RoBERTa, consist of millions or even billions of parameters. Training these models involves performing numerous mathematical operations, such as matrix multiplications and convolutions, on these parameters. CUDA enables these computations to be parallelized across the thousands of cores present in a GPU, greatly speeding up the training process compared to a traditional CPU.

- **Optimized tensor operations for PyTorch or TensorFlow**, which are designed to work seamlessly with CUDA. These frameworks provide GPU-accelerated libraries that implement optimized tensor operations specifically designed for GPUs. Tensors are multi-dimensional arrays used to store and manipulate data in DL models. With CUDA, these tensor operations can be efficiently executed on the GPU, leveraging its high memory bandwidth and parallel processing capabilities.

- **High memory bandwidth**, which enables data to be transferred to and from the GPU memory at a much faster rate, enabling faster data processing during training. DL models often require large amounts of data to be loaded and processed in batches. CUDA allows these batches to be efficiently transferred and processed on the GPU, reducing training time.

By utilizing CUDA, DL frameworks can effectively leverage the parallel computing capabilities and optimized operations of NVIDIA GPUs, resulting in significant acceleration of the training process for large-scale models such as RoBERTa.

Training and evaluation of an AE

We mentioned AEs in *Chapter 7* when we discussed the process of feature engineering for images. AEs, however, are used to do much more than just image feature extraction. One of the major aspects of them is to be able to recreate images. This means that we can create images based on the placement of the image in the latent space.

So, let us train the AE model for a dataset that is pretty standard in ML – Fashion MNIST. We got to see what the dataset looks like in our previous chapters. We start our training by preparing the data in the following code fragment:

```
# Transforms images to a PyTorch Tensor
tensor_transform = transforms.ToTensor()

# Download the Fashion MNIST Dataset
dataset = datasets.FashionMNIST(root = "./data",
                                train = True,
                                download = True,
                                transform = tensor_transform)

# DataLoader is used to load the dataset
# for training
loader = torch.utils.data.DataLoader(dataset = dataset,
                                     batch_size = 32,
                                     shuffle = True)
```

It imports the necessary modules from the PyTorch library.

It defines a transformation called `tensor_transform` using `transforms.ToTensor()`. This transformation is used to convert images in the dataset to PyTorch tensors.

The code fragment downloads the dataset using the `datasets.FashionMNIST()` function. The `train` parameter is set to `True` to indicate that the downloaded dataset is for training purposes. The `download` parameter is set to `True` to automatically download the dataset if it is not already present in the specified directory.

Since we use the PyTorch framework with accelerated computing, we need to make sure that the image is transformed into a tensor. The `transform` parameter is set to `tensor_transform`, which is a transformer defined in the first line of the code fragment.

Then, we create a `DataLoader` object used to load the dataset in batches for training. The `dataset` parameter is set to the previously downloaded dataset. The `batch_size` parameter is set to `32`, indicating that each batch of the dataset will contain 32 images.

The `shuffle` parameter is set to `True` to shuffle the order of the samples in each epoch of training, ensuring randomization and reducing any potential bias during training.

Once we have the dataset prepared, we can create our AE, which we do like this:

```
# Creating a PyTorch class
# 28*28 ==> 9 ==> 28*28
class AE(torch.nn.Module):
    def __init__(self):
        super().__init__()

        # Building an linear encoder with Linear
        # layer followed by Relu activation function
        # 784 ==> 9
        self.encoder = torch.nn.Sequential(
            torch.nn.Linear(28 * 28, 128),
            torch.nn.ReLU(),
            torch.nn.Linear(128, 64),
            torch.nn.ReLU(),
            torch.nn.Linear(64, 36),
            torch.nn.ReLU(),
            torch.nn.Linear(36, 18),
            torch.nn.ReLU(),
            torch.nn.Linear(18, 9)
        )

        # Building an linear decoder with Linear
        # layer followed by Relu activation function
        # The Sigmoid activation function
        # outputs the value between 0 and 1
        # 9 ==> 784
        self.decoder = torch.nn.Sequential(
            torch.nn.Linear(9, 18),
            torch.nn.ReLU(),
            torch.nn.Linear(18, 36),
            torch.nn.ReLU(),
            torch.nn.Linear(36, 64),
            torch.nn.ReLU(),
            torch.nn.Linear(64, 128),
            torch.nn.ReLU(),
            torch.nn.Linear(128, 28 * 28),
            torch.nn.Sigmoid()
        )

    def forward(self, x):
        encoded = self.encoder(x)
        decoded = self.decoder(encoded)
        return decoded
```

First, we define a class named AE that inherits from the `torch.nn.Module` class, which is the base class for all NN modules in PyTorch. The `super().__init__()` line ensures that the initialization of the base class (`torch.nn.Module`) is called. Since AEs are a special kind of NN class with backpropagation learning, we can just inherit a lot of basic functionality from the library.

Then, we define the encoder part of the AE. The encoder consists of several linear (fully connected) layers with ReLU activation functions. Each `torch.nn.Linear` layer represents a linear transformation of the input data followed by an activation function. In this case, the input size is 28 * 28 (which corresponds to the dimensions of an image in the Fashion MNIST dataset), and the output size gradually decreases until it reaches 9, which is our latent vector size.

Then, we define the decoder part of the AE. The decoder is responsible for reconstructing the input data from the encoded representation. It consists of several linear layers with ReLU activation functions, followed by a final linear layer with a sigmoid activation function. The input size of the decoder is 9, which corresponds to the size of the latent vector space in the bottleneck of the encoder. The output size is 28 * 28, which matches the dimensions of the original input data.

The `forward` method defines the forward pass of the AE. It takes an x input and passes it through the encoder to obtain an encoded representation. Then, it passes the encoded representation through the decoder to reconstruct the input data. The reconstructed output is returned as the result. We are now ready to instantiate our AE:

```
# Model Initialization
model = AE()

# Validation using MSE Loss function
loss_function = torch.nn.MSELoss()

# Using an Adam Optimizer with lr = 0.1
optimizer = torch.optim.Adam(model.parameters(),
                             lr = 1e-1,
                             weight_decay = 1e-8)
```

In this code, we first instantiate our AE as our model. Then, we create an instance of the **Mean Squared Error** (**MSE**) loss function provided by PyTorch. MSE is a commonly used loss function for regression tasks. We need it to calculate the mean squared difference between the predicted values and the target values – which are the individual pixels in our dataset, providing a measure of how well the model is performing. *Figure 11.4* shows the role of the learning function in the process of training the AE:

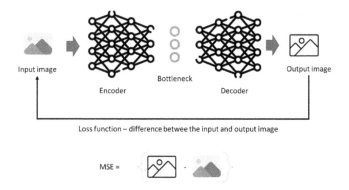

Figure 11.4 – Loss function (MSE) in the AE training process

Then, we initialize the optimizer used to update the model's parameters during training. In this case, the code creates an Adam optimizer, which is a popular optimization algorithm for training NNs. It takes three important arguments:

- `model.parameters()`: This specifies the parameters that will be optimized. In this case, it includes all the parameters of the model (the AE) that we created earlier.

- `lr=1e-1`: This sets the learning rate, which determines the step size at which the optimizer updates the parameters. A higher learning rate can lead to faster convergence but may risk overshooting the optimal solution, while a lower learning rate may converge more slowly but with potentially better accuracy.

- `weight_decay=1e-8`: This parameter adds a weight decay regularization term to the optimizer. Weight decay helps prevent overfitting by adding a penalty term to the loss function that discourages large weights. The `1e-8` value represents the weight decay coefficient.

With this code, we have now an instance of an AE to train. Now, we can start the process of training. We train the model for 10 epochs, but we can try more if needed:

```
epochs = 10
outputs = []
losses = []
for epoch in range(epochs):
    for (image, _) in loader:

        # Reshaping the image to (-1, 784)
        image = image.reshape(-1, 28*28)

        # Output of Autoencoder
        reconstructed = model(image)

        # Calculating the loss function
```

```
        loss = loss_function(reconstructed, image)

        # The gradients are set to zero,
        # the gradient is computed and stored.
        # .step() performs parameter update
        optimizer.zero_grad()
        loss.backward()
        optimizer.step()

        # Storing the losses in a list for plotting
        losses.append(loss)
    outputs.append((epochs, image, reconstructed))
```

We start by iterating over the specified number of epochs for the training. Within each epoch, we iterate over the loader, which provides batches of image data and their corresponding labels. We do not use the labels, because the AE is a network to recreate images and not to learn what the images show – in that sense, it is an unsupervised model.

For each image, we reshape the input image data by flattening each image, originally in the shape of (batch_size, 28, 28), into a 2D tensor of shape (batch_size, 784), where each row represents a flattened image. The flattened image is created when we take each row of pixels and concatenate it to create one large vector. It is needed as the images are two-dimensional, while our tensor input needs to be of a single dimension.

Then, we obtain the reconstructed image using reconstructed = model(image). Once we get the reconstructed image, we can calculate the MSE loss function and use that information to manage the next step of the learning (optimizer.zero_grad()). In the last line, we add this information to the list of losses per iteration so that we can create a learning diagram. We do it by using the following code fragment:

```
# Defining the Plot Style
plt.style.use('seaborn')
plt.xlabel('Iterations')
plt.ylabel('Loss')

# Convert the list to a PyTorch tensor
losses_tensor = torch.tensor(losses)

plt.plot(losses_tensor.detach().numpy()[::-1])
```

This results in a learning diagram, shown in *Figure 11.5*:

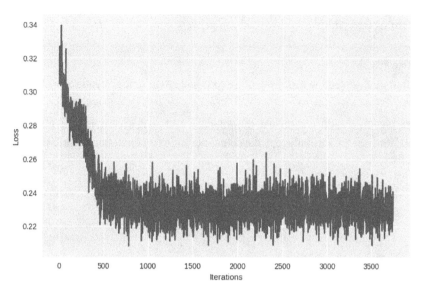

Figure 11.5 – Learning rate diagram from training our AE

The learning rate diagram shows that the AE is not really great yet and that we should train it a bit more. However, we can always check what the recreated images look like. We can do that using this code:

```
for i, item in enumerate(image):

    # Reshape the array for plotting
    item = item.reshape(-1, 28, 28)
    plt.imshow(item[0])
```

The code results in the output shown in *Figure 11.6*:

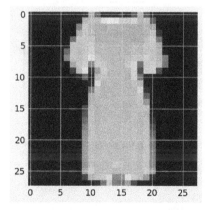

Figure 11.6 – Recreated image from our AE

Despite the learning rate, which is OK, we still can get very good results from our AEs.

> **Best practice #58**
>
> In addition to monitoring the loss, make sure to visualize the actual results of the generation.

Monitoring the loss function is a good way to understand when the AE stabilizes. However, just the loss function is not enough. I usually plot the actual output to understand whether the AE has been trained correctly.

Finally, we can visualize the learning process when we use this code:

```
yhat = model(image[0])

make_dot(yhat,
         params=dict(list(model.named_parameters())),
         show_attrs=True,
         show_saved=True)
```

This code visualizes the learning process of the entire network. It creates a large image, and we can only show a small excerpt of it. *Figure 11.7* shows this excerpt:

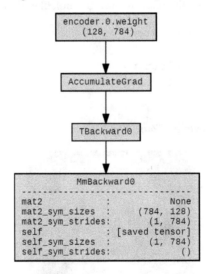

Figure 11.7 – The first three steps in training an AE, visualized as the AE architecture

We can even visualize the entire architecture in a text form by using the following code:

```
from torchsummary import summary
summary(model, (1, 28 * 28))
```

This results in the following model:

```
-----------------------------------------------------------------
        Layer (type)              Output Shape           Param #
=================================================================
        Linear-1                  [-1, 1, 128]           100,480
          ReLU-2                  [-1, 1, 128]                 0
        Linear-3                  [-1, 1,  64]             8,256
          ReLU-4                  [-1, 1,  64]                 0
        Linear-5                  [-1, 1,  36]             2,340
          ReLU-6                  [-1, 1,  36]                 0
        Linear-7                  [-1, 1,  18]               666
          ReLU-8                  [-1, 1,  18]                 0
        Linear-9                  [-1, 1,   9]               171
       Linear-10                  [-1, 1,  18]               180
         ReLU-11                  [-1, 1,  18]                 0
       Linear-12                  [-1, 1,  36]               684
         ReLU-13                  [-1, 1,  36]                 0
       Linear-14                  [-1, 1,  64]             2,368
         ReLU-15                  [-1, 1,  64]                 0
       Linear-16                  [-1, 1, 128]             8,320
         ReLU-17                  [-1, 1, 128]                 0
       Linear-18                  [-1, 1, 784]           101,136
      Sigmoid-19                  [-1, 1, 784]                 0
=================================================================
Total params: 224,601
Trainable params: 224,601
Non-trainable params: 0
-----------------------------------------------------------------
Input size (MB): 0.00
Forward/backward pass size (MB): 0.02
Params size (MB): 0.86
Estimated Total Size (MB): 0.88
-----------------------------------------------------------------
```

The bottleneck layer is in boldface, to illustrate the place where the encode and decode parts are linked to one another.

Developing safety cages to prevent models from breaking the entire system

As GenAI systems such as MLMs and AEs create new content, there is a risk that they generate content that can either break the entire software system or become unethical.

Therefore, software engineers often use the concept of a safety cage to guard the model itself from inappropriate input and output. For an MLM such as RoBERTa, this can be a simple preprocessor that checks whether the content generated is problematic. Conceptually, this is illustrated in *Figure 11.8*:

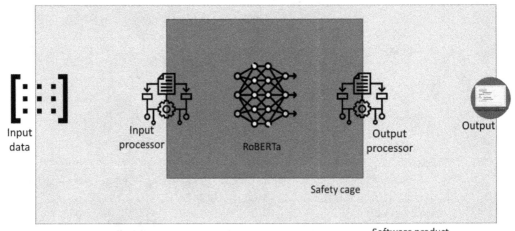

Figure 11.8 – Safety-cage concept for MLMs

In the example of the wolfBERTa model, this can mean that we check whether the generated code does not contain cybersecurity vulnerabilities, which can potentially allow hackers to take over our system. This means that all programs generated by the wolfBERTa model should be checked using tools such as SonarQube or CodeSonar to check for cybersecurity vulnerabilities, hence my next best practice.

> **Best practice #59**
>
> Check the output of GenAI models so that it does not break the entire system or provide unethical responses.

My recommendation to create such safety cages is to start from the requirements of the system. The first step is to understand what the system is going to do and understand which dangers and risks this task entails. The safety cage's output processor should ensure that these dangerous situations do not occur and that they are handled properly.

Once we understand how to prevent dangers, we can move over to conceptualizing how to prevent these risks on the language-model level. For example, when we train the model, we can select code that is known to be secure and does not contain security vulnerabilities. Although it does not guarantee that the model generates secure code, it certainly reduces the risk for it.

Summary

In this chapter, we learned how to train advanced models and saw that their training is not much more difficult than training classical ML models, which were described in *Chapter 10*. Even though the models that we trained are much more complex than the models in *Chapter 10*, we can use the same principles and expand this kind of activity to train even more complex models.

We focused on GenAI in the form of BERT models (fundamental GPT models) and AEs. Training these models is not very difficult, and we do not need huge computing power to train them. Our wolfBERTa model has ca. 80 million parameters, which seems like a lot, but the really good models, such as GPT-3, have billions of parameters – GPT-3 has 175 billion parameters, NVIDIA Turing has over 350 billion parameters, and GPT-4 is 1,000 times larger than GPT-3. The training process is the same, but we need a supercomputing architecture in order to train these models.

We have also learned that these models are only parts of larger software systems. In the next chapter, we learn how to create such a larger system.

References

- *Kratsch, W. et al., Machine learning in business process monitoring: a comparison of deep learning and classical approaches used for outcome prediction. Business & Information Systems Engineering, 2021, 63: p. 261-276.*

- *Vaswani, A. et al., Attention is all you need. Advances in neural information processing systems, 2017, 30.*

- *Aggarwal, A., M. Mittal, and G. Battineni, Generative adversarial network: An overview of theory and applications. International Journal of Information Management Data Insights, 2021. 1(1): p. 100004.*

- *Creswell, A., et al., Generative adversarial networks: An overview. IEEE signal processing magazine, 2018. 35(1): p. 53-65.*

12

Designing Machine Learning Pipelines (MLOps) and Their Testing

MLOps, short for machine learning (ML) operations, is a set of practices and techniques aimed at streamlining the deployment, management, and monitoring of ML models in production environments. It borrows concepts from the DevOps (development and operations) approach, adapting them to the unique challenges posed by ML.

The main goal of MLOps is to bridge the gap between data science and operations teams, fostering collaboration and ensuring that ML projects can be effectively and reliably deployed at scale. MLOps helps to automate and optimize the entire ML life cycle, from model development to deployment and maintenance, thus improving the efficiency and effectiveness of ML systems in production.

In this chapter, we learn how ML systems are designed and operated in practice. The chapter shows how pipelines are turned into a software system, with a focus on testing ML pipelines and their deployment at Hugging Face.

In this chapter, we're going to cover the following main topics:

- What ML pipelines are
- ML pipelines – how to use ML in the system in practice
- Raw data-based pipelines
- Feature-based pipelines
- Testing of ML pipelines
- Monitoring ML systems at runtime

What ML pipelines are

Undoubtedly, in recent years, the field of ML has witnessed remarkable advancements, revolutionizing industries and empowering innovative applications. As the demand for more sophisticated and accurate models grows, so does the complexity of developing and deploying them effectively. The industrial introduction of ML systems called for more rigorous testing and validation of these ML-based systems. In response to these challenges, the concept of ML pipelines has emerged as a crucial framework to streamline the entire ML development process, from data preprocessing and feature engineering to model training and deployment. This chapter explores the applications of MLOps in the context of both cutting-edge **deep learning** (**DL**) models such as **Generative Pre-trained Transformer** (**GPT**) and traditional classical ML models.

We begin by exploring the underlying concepts of ML pipelines, stressing their importance in organizing the ML workflow and promoting collaboration between data scientists and engineers. We synthesize a lot of knowledge presented in the previous chapters – data quality assessment, model inference, and monitoring.

Next, we discuss the unique characteristics and considerations involved in building pipelines for GPT models and similar, leveraging their pre-trained nature to tackle a wide range of language tasks. We explore the intricacies of fine-tuning GPT models on domain-specific data and the challenges of incorporating them into production systems.

Following the exploration of GPT pipelines, we shift our focus to classical ML models, examining the feature engineering process and its role in extracting relevant information from raw data. We delve into the diverse landscape of traditional ML algorithms, understanding when to use each approach, and their trade-offs in different scenarios.

Finally, we show how to test ML pipelines, and we emphasize the significance of model evaluation and validation in assessing performance and ensuring robustness in production environments. Additionally, we examine strategies for model monitoring and maintenance, safeguarding against concept drift and guaranteeing continuous performance improvement.

ML pipelines

An ML pipeline is a systematic and automated process that organizes the various stages of an ML workflow. It encompasses the steps involved in preparing data, training an ML model, evaluating its performance, and deploying it for use in real-world applications. The primary goal of an ML pipeline is to streamline the end-to-end ML process, making it more efficient, reproducible, and scalable.

An ML pipeline typically consists of the following essential components:

- **Data collection, preprocessing, and wrangling**: In this initial stage, relevant data is gathered from various sources and prepared for model training. Data preprocessing involves cleaning, transforming, and normalizing the data to ensure it is in a suitable format for the ML algorithm.

- **Feature engineering and selection**: Feature engineering involves selecting and creating relevant features (input variables) from the raw data that will help the model learn patterns and make accurate predictions. Proper feature selection is crucial in improving model performance and reducing computational overhead.

- **Model selection and training**: In this stage, one or more ML algorithms are chosen, and the model is trained on the prepared data. Model training involves learning underlying patterns and relationships in the data to make predictions or classifications.

- **Model evaluation and validation**: The trained model is evaluated using metrics such as accuracy, precision, recall, F1-score, and so on, to assess its performance on unseen data. Cross-validation techniques are often used to ensure the model's generalization capability.

- **Hyperparameter tuning**: Many ML algorithms have hyperparameters, which are adjustable parameters that control the model's behavior. Hyperparameter tuning involves finding the optimal values for these parameters to improve the model's performance.

- **Model deployment**: Once the model has been trained and validated, it is deployed into a production environment, where it can make predictions on new, unseen data. Model deployment may involve integrating the model into existing applications or systems.

- **Model monitoring and maintenance**: After deployment, the model's performance is continuously monitored to detect any issues or drift in performance. Regular maintenance may involve retraining the model with new data to ensure it remains accurate and up to date.

An ML pipeline provides a structured framework for managing the complexity of ML projects, enabling data scientists and engineers to collaborate more effectively and ensuring that models can be developed and deployed reliably and efficiently. It promotes reproducibility, scalability, and ease of experimentation, facilitating the development of high-quality ML solutions. *Figure 12.1* shows a conceptual model of an ML pipeline, which we introduced in *Chapter 2*:

Figure 12.1 – ML pipeline: a conceptual overview

We covered the elements of the blue-shaded elements in the previous chapters, and here, we focus mostly on the parts that are not covered yet. However, before we dive into the technical elements of this pipeline, let us introduce the concept of MLOps.

Elements of MLOps

As the main goal of MLOps is to bridge the gap between data science and operations teams, MLOps automates and optimizes the entire ML life cycle, from model development to deployment and maintenance, thus improving the efficiency and effectiveness of ML systems in production.

Key components and practices in MLOps include:

- **Version control**: Applying **version control systems (VCSs)** such as Git to manage and track changes in ML code, datasets, and model versions. This enables easy collaboration, reproducibility, and tracking of model improvements.

- **Continuous integration and continuous deployment (CI/CD)**: Leveraging CI/CD pipelines to automate the testing, integration, and deployment of ML models. This helps ensure that changes to the code base are seamlessly deployed to production while maintaining high-quality standards.

- **Model packaging**: Creating standardized, reproducible, and shareable containers or packages for ML models, making it easier to deploy them across different environments consistently.

- **Model monitoring**: Implementing monitoring and logging solutions to keep track of the model's performance and behavior in real time. This helps detect issues early and ensure the model's ongoing reliability.

- **Scalability and infrastructure management**: Designing and managing the underlying infrastructure to support the demands of the ML models in production, ensuring they can handle increased workloads and scale efficiently.

- **Model governance and compliance**: Implementing processes and tools to ensure compliance with legal and ethical requirements, privacy regulations, and company policies when deploying and using ML models.

- **Collaboration and communication**: Facilitating effective communication and collaboration between data scientists, engineers, and other stakeholders involved in the ML deployment process.

By adopting MLOps principles, organizations can accelerate the development and deployment of ML models while maintaining their reliability and effectiveness in real-world applications. It also helps reduce the risk of deployment failures and promotes a culture of collaboration and continuous improvement within data science and operations teams.

ML pipelines – how to use ML in the system in practice

Training and validating ML models on a local platform is the beginning of the process of using an ML pipeline. After all, it would be of limited use if we had to retrain the ML models on every computer from our customers.

Therefore, we often deploy ML models to a model repository. There are a few popular ones, but the one that is used by the largest community is the HuggingFace repository. In that repository, we can deploy both the models and datasets and even create spaces where the models can be used for experiments without the need to download them. Let us deploy the model trained in *Chapter 11* to that repository. For that, we need to have an account at `huggingface.com`, and then we can start.

Deploying models to HuggingFace

First, we need to create a new model using the **New** button on the main page, as in *Figure 12.2*:

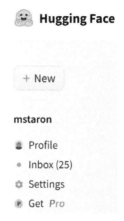

Figure 12.2 – New button to create a model

Then, we fill in information about our model to create space for it. *Figure 12.3* presents a screenshot of this process. In the form, we fill in the name of the model, whether it should be private or public, and we choose a license for it. In this example, we go with the MIT License, which is very permissive and allows everyone to use, reuse, and redistribute the model as long as they include the MIT License text along with it:

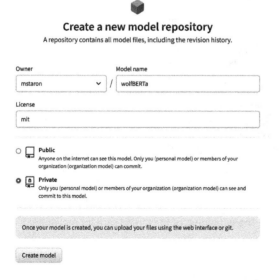

Figure 12.3 – Model metadata card

Once the model has been created, we get a space where we can start deploying the model. The empty space looks like the one in *Figure 12.4*:

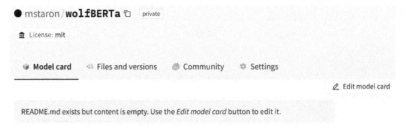

Figure 12.4 – Empty model space

The top menu contains four options, but the first two are the most important ones – **Model card** and **Files and versions**. The model card is a short description of the model. It can contain any kind of information, but the most common information is how to use the model. We follow this convention and prepare the model card as shown in *Figure 12.5*:

Figure 12.5 – The beginning of the model card for our wolfBERTa model

> **Best practice #60**
>
> The model card should contain information about how the model was trained, how to use it, which tasks it supports, and how to reference the model.

Since HuggingFace is a community, it is important to properly document models created and provide information on how the models were trained and what they can do. Therefore, my best practice is to include all that information in the model card. Many models include also information about how to contact the authors and whether the models had been pre-trained before they were trained.

Once the model card is ready, we can move to the **Files and versions** section of the model space. In that space, we can see files that have been created so far (that is, `Readme.txt` – the model card), and we can add actual model files (see *Figure 12.6*):

Figure 12.6 – Files and versions of models; we can add a model by
using the Add file button in the top right-hand corner

Once we click on the **Add file** button, we can add model files. We find the model files in the same repository that we used in *Chapter 11*, in the `wolfBERTa` subfolder. That folder contains the following files:

```
Mode                LastWriteTime                      Name
------              -------------                      ----
d----l          2023-07-01      10:25          checkpoint-340000
d----l          2023-07-01      10:25          checkpoint-350000
-a---l          2023-06-27      21:30          config.json
-a---l          2023-06-27      17:55          merges.txt
-a---l          2023-06-27      21:30          pytorch_model.bin
-a---l          2023-06-27      21:30          training_args.bin
-a---l          2023-06-27      17:55          vocab.json
```

The first two entries are the model checkpoints; that is, the versions of the model saved during our training process. These two folders are not important for the deployment, and therefore they will be ignored. The rest of the files should be copied to the newly created model repository at HuggingFace.

The model, after uploading, should look something like the one presented in *Figure 12.7*:

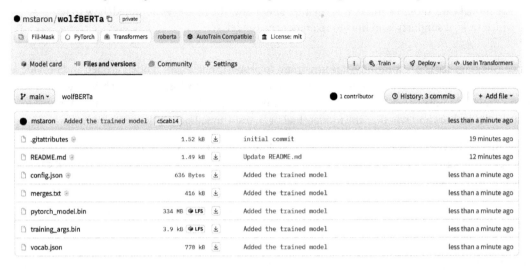

Figure 12.7 – Model uploaded to the HuggingFace repository

After this, the model is ready to be used by the community. What we can also do is create an inference API for the community to quickly test our models. It is provided to us automatically once we go back to the **Model card** menu, under the **Hosted inference API** section (right-hand side of *Figure 12.8*):

Figure 12.8 – Hosted inference API provided automatically for our model

When we input `int HTTP_get(<mask>)`, we ask the model to provide the input parameter for that function. The results show that the most probable token is `void` and the second in line is the `int` token. Both are relevant as they are types used in parameters, but they are probably not going to make this program compile, so we would need to develop a loop that would predict more than just one token for the program. It probably needs a bit more training as well.

Now, we have a fully deployed model that can be used in other applications without much hassle.

Downloading models from HuggingFace

We have already seen how to download a model from HuggingFace, but for the sake of completeness, let's see how this is done for the `wolfBERTa` model. Essentially, we follow the model card and use the following Python code fragment:

```python
from transformers import pipeline
unmasker = pipeline('fill-mask', model='mstaron/wolfBERTa')
unmasker("Hello I'm a <mask> model.")
```

This code fragment downloads the model and uses an `unmasker` interface to make an inference using the `fill-mask` pipeline. The pipeline allows you to input a sentence with a `<mask>` masked token, and the model will attempt to predict the most suitable word to fill in the masked position. The three lines of this code fragment do the following:

- `from transformers import pipeline`: This line imports the pipeline function from the `transformers` library. The pipeline function simplifies the process of using pre-trained models for various **natural language processing** (**NLP**) tasks.

- `unmasker = pipeline('fill-mask', model='mstaron/wolfBERTa')`: This line creates a new pipeline named `unmasker` for the task. The pipeline will use the pre-trained `wolfBERTa` model.

- `unmasker("Hello I'm a <mask> model.")`: This line utilizes the `unmasker` pipeline to predict the word that best fits the masked position in the given sentence. The `<mask>` token indicates the position where the model should try to fill in a word.

When this line is executed, the pipeline will call the `wolfBERTa` model, which will make predictions based on the provided sentence. The model will predict the word that it finds to best complete the sentence in the position of the `<mask>` token.

One can use other models in a very similar way. The main advantage of a community model hub such as HuggingFace is that it provides a great way to uniformly manage models and pipelines and allows us to quickly exchange models in software products.

Raw data-based pipelines

Creating a full pipeline can be a daunting task and requires creating customized tools for all models and all kinds of data. It allows us to optimize how we use the models, but it requires a lot of effort. The main rationale behind pipelines is that they link two areas of ML – the model and its computational capabilities with the task and the data from the domain. Luckily for us, the main model hubs such as HuggingFace have an API that provides ML pipelines automatically. Pipelines in HuggingFace are related to the model and provided by the framework based on the model's architecture, input, and output.

Pipelines for NLP-related tasks

Text classification is a pipeline designed to classify text input into predefined categories or classes. It's particularly useful for tasks such as **sentiment analysis (SA)**, topic categorization, spam detection, intent recognition, and so on. The pipeline typically employs pre-trained models fine-tuned on specific datasets for different classification tasks. We have seen similar capabilities in *Part I* of this book when we used ML for SA of code reviews.

An example is presented in the following code fragment:

```
from transformers import pipeline

# Load the text classification pipeline
classifier = pipeline("text-classification")

# Classify a sample text
result = classifier("This movie is amazing and highly recommended!")
print(result)
```

The code fragment shows that there are essentially two lines of code (in boldface) that we need to instantiate the pipeline, as we've also seen before.

Text generation is another pipeline that allows the generating of text using pre-trained language models, such as GPT-3, based on a provided prompt or seed text. It's capable of generating human-like text for various applications, such as chatbots, creative writing, **question answering (QA)**, and more.

An example is presented in the following code fragment:

```
from transformers import pipeline

# Load the text generation pipeline
generator = pipeline("text-generation")

# Generate text based on a prompt
prompt = "In a galaxy far, far away… "
result = generator(prompt, max_length=50, num_return_sequences=3)
for output in result:
    print(output['generated_text'])
```

Summarization is a pipeline designed to summarize longer texts into shorter, coherent summaries. It utilizes transformer-based models that have been trained on large datasets with a focus on the summarization task. The pipeline is exemplified in the following code fragment:

```
from transformers import pipeline

# Load the summarization pipeline
summarizer = pipeline("summarization")

# Summarize a long article
article = """
In a groundbreaking discovery, scientists have found a new species
of dinosaur in South America. The dinosaur, named "Titanus maximus,"
is estimated to have been the largest terrestrial creature to ever
walk the Earth. It belonged to the sauropod group of dinosaurs, known
for their long necks and tails. The discovery sheds new light on the
diversity of dinosaurs that once inhabited our planet.
"""
result = summarizer(article, max_length=100, min_length=30, do_
sample=False)
print(result[0]['summary_text'])
```

There are more pipelines in the HuggingFace `transformers` API, so I encourage you to take a look at these pipelines. However, my best practice related to pipelines is this:

> **Best practice #61**
>
> Experiment with different models to find the best pipeline.

Since the API provides the same pipeline for similar models, changing the model or its version is quite simple. Therefore, we can create a product based on a model that has similar (but not the same) capabilities as the model that we use and simultaneously train the model.

Pipelines for images

Pipelines for image processing are designed specifically for tasks related to image processing. The HuggingFace hub contains several of these pipelines, with the following ones being the most popular.

Image classification is designed specifically to classify an image to a specific class. It is the same kind of task as is probably the most widely known – classifying an image to be "cat", "dog", or "car". The following code example (from the HuggingFace tutorial) shows the usage of an image classification pipeline:

```
from transformers import pipeline

# first, create an instance of the image classification pipeline for
the selected model
```

```
classifier = pipeline(model="microsoft/beit-base-patch16-224-pt22k-
ft22k")

# now, use the pipeline to classify an image
classifier("https://huggingface.co/datasets/Narsil/image_dummy/raw/
main/parrots.png")
```

The preceding code fragment shows that an image classification pipeline is created equally easily (if not easier) as pipelines for text analysis tasks.

An image segmentation pipeline is used when we want to add a so-called semantic map to an image (see *Figure 12.9*):

Figure 12.9 – Semantic map of an image, the same as we saw in Chapter 3

An example code fragment that contains such a pipeline is presented next (again, from the HuggingFace tutorial):

```
from transformers import pipeline

segmenter = pipeline(model="facebook/detr-resnet-50-panoptic")
segments = segmenter("https://huggingface.co/datasets/Narsil/image_
dummy/raw/main/parrots.png")
segments[0]["label"]
```

The preceding code fragment creates an image segmentation pipeline, uses it, and stores the results in a `segments` list. The last line of the list prints the label of the first segment. Using the `segments[0]` `["mask"].size` statement, we can receive the size of the image map in pixels.

An object detection pipeline is used for tasks that require the recognition of objects of a predefined class in the image. We have seen an example of this task in *Chapter 3* already. The code for this kind of pipeline looks very similar to the previous ones:

```
from transformers import pipeline

detector = pipeline(model="facebook/detr-resnet-50")
```

```
detector("https://huggingface.co/datasets/Narsil/image_dummy/raw/main/
parrots.png")
```

Executing this code creates a list of bounding boxes of objects detected in the image, together with its bounding boxes. My best practices related to the use of pipelines for images are the same as for language tasks.

Feature-based pipelines

Feature-based pipelines do not have specific classes because they are much lower level. They are the `model.fit()` and `model.predict()` statements from the standard Python ML implementation. These pipelines require software developers to prepare the data manually and also to take care of the results manually; that is, by implementing preprocessing steps such as converting data to tables using one-hot encoding and post-processing steps such as converting the data into a human-readable output.

An example of this kind of pipeline was the prediction of defects that we have seen in the previous parts of the book; therefore, they do not need to be repeated.

What is important, however, is that all pipelines are the way that link the ML domain with the software engineering domain. The first activity that I do after developing a pipeline is to test it.

Testing of ML pipelines

Testing of ML pipelines is done at multiple levels, starting with unit tests and moving up toward integration (component) tests and then to system and acceptance tests. In these tests, two elements are important – the model itself and the data (for the model and the oracle).

Although we can use the unit test framework included in Python, I strongly recommend using the Pytest framework instead, due to its simplicity and flexibility. We can install this framework by simply using this command:

```
>> pip install pytest
```

That will download and install the required packages.

> **Best practice #62**
> Use a professional testing framework such as Pytest.

Using a professional framework provides us with the compatibility required by MLOps principles. We can share our models, data, source code, and all other elements without the need for cumbersome setup and installation of the frameworks themselves. For Python, I recommend using the Pytest framework as it is well known, widely used, and supported by a large community.

Here is a code fragment that downloads a model and prepares it for being tested:

```
# import json to be able to read the embedding vector for the test
import json

# import the model via the huggingface library
from transformers import AutoTokenizer, AutoModelForMaskedLM

# load the tokenizer and the model for the pretrained SingBERTa
tokenizer = AutoTokenizer.from_pretrained('mstaron/SingBERTa')

# load the model
model = AutoModelForMaskedLM.from_pretrained("mstaron/SingBERTa")

# import the feature extraction pipeline
from transformers import pipeline

# create the pipeline, which will extract the embedding vectors
# the models are already pre-defined, so we do not need to train
anything here
features = pipeline(
    "feature-extraction",
    model=model,
    tokenizer=tokenizer,
    return_tensor = False
)
```

This code snippet is used to load and set up a pre-trained language model, specifically the SingBERTa model, using the Hugging Face transformers library. It contains the following elements:

1. Import the necessary modules from the transformers library:

    A. AutoTokenizer: This class is used to automatically select the appropriate tokenizer for the pre-trained model.

    B. AutoModelForMaskedLM: This class is used to automatically select the appropriate model for **masked language modeling (MLM)** tasks.

2. Load the tokenizer and model for the pre-trained SingBERTa model:

    A. tokenizer = AutoTokenizer.from_pretrained('mstaron/SingBERTa'): This line loads the tokenizer for the pre-trained SingBERTa model from the Hugging Face model hub.

    B. model = AutoModelForMaskedLM.from_pretrained("mstaron/SingBERTa"): This line loads the pre-trained SingBERTa model.

3. Import the feature extraction pipeline:

    A. `from transformers import pipeline`: This line imports the pipeline class from the `transformers` library, which allows us to easily create pipelines for various NLP tasks.

4. Create a feature extraction pipeline:

    A. `features = pipeline("feature-extraction", model=model, tokenizer=tokenizer, return_tensor=False)`: This line creates a pipeline for feature extraction. The pipeline uses the pre-trained model and tokenizer loaded earlier to extract embedding vectors from the input text. The `return_tensor=False` argument ensures that the output will be in a non-tensor format (likely NumPy arrays or Python lists).

With this setup, you can now use the `features` pipeline to extract embedding vectors from text input using the pre-trained `SingBERTa` model without the need for any additional training. We've seen this model being used before, so here, let us focus on its testing. The following code fragment is a test case to check that the model has been downloaded correctly and that it is ready to be used:

```
def test_features():

    # get the embeddings of the word "Test"
    lstFeatures = features("Test")

    # read the oracle from the json file
    with open('test.json', 'r') as f:
        lstEmbeddings = json.load(f)

    # assert the embeddings and the oracle are the same
    assert lstFeatures[0][0] == lstEmbeddings
```

This code fragment defines a `test_features()` test function. The purpose of this function is to test the correctness of the feature extraction pipeline created in the previous code snippet by comparing the embeddings of the word `"Test"` obtained from the pipeline to the expected embeddings stored in a JSON file named `'test.json'`. The content of that file is our oracle, and it is a large vector of numbers that we use to compare to the actual model output:

- `lstFeatures = features("Test")`: This line uses the previously defined `features` pipeline to extract embeddings for the word `"Test"`. The `features` pipeline was created using the pre-trained `SingBERTa` model and tokenizer. The pipeline takes the input `"Test"`, processes it through the tokenizer, passes it through the model, and returns embedding vectors as `lstFeatures`.

- `with open('test.json', 'r') as f:`: This line opens the `'test.json'` file in read mode using a context manager (`with` statement).

- `lstEmbeddings = json.load(f):` This line reads the contents of the `'test.json'` file and loads its content into the `lstEmbeddings` variable. The JSON file should contain a list of embedding vectors representing the expected embeddings for the word `"Test"`.

- `assert lstFeatures[0][0] == lstEmbeddings:` This line performs an assertion to check if the embedding vector obtained from the pipeline (`lstFeatures[0][0]`) is equal to the expected embedding vector (oracle) from the JSON file (`lstEmbeddings`). A comparison is made by checking whether the elements at the same position in both lists are the same.

If the assertion is `true` (that is, the pipeline's extracted embedding vector is the same as the expected vector from the JSON file), the test will pass without any output. However, if the assertion is `false` (that is, the embeddings do not match), the test framework (Pytest) marks this test case as failed.

In order to execute the tests, we can write the following statement in the same directory as our project:

```
>> pytest
```

In our case, this results in the following output (redacted for brevity):

```
==================== test session starts ====================
platform win32 -- Python 3.11.4, pytest-7.4.0, pluggy-1.2.0
rootdir: C:\machine_learning_best_practices\chapter_12
plugins: anyio-3.7.0
collected 1 item

chapter_12_download_model_test.py .                    [100%]

===================== 1 passed in 4.17s ====================
```

This fragment shows that the framework found one test case (`collected 1 item`) and that it executed it. It also says that the test case passed in 4.17 seconds.

Therefore, here comes my next best practice.

> **Best practice #63**
> Set up your test infrastructure based on your training data.

Since the models are inherently probabilistic, it is best to test the models based on the training data. Here, I do not mean that we test the performance in the sense of ML, like accuracy. I mean that we test that the models actually work. By using the same data as we used for the training, we can check whether the models' inference is correct for the data that we used before. Therefore, I mean this as testing in the software engineering sense of this term.

Now, analogous to the language models presented previously, we can use a similar approach to test a classical ML model. It's sometimes called a zero-table test. In this test, we use simple data with one data point only to test that the model's predictions are correct. Here is how we set up such a test:

```
# import the libraries pandas and joblib
import pandas as pd
import joblib

# load the model
model = joblib.load('./chapter_12_decision_tree_model.joblib')

# load the data that we used for training
dfDataAnt13 = pd.read_excel('./chapter_12.xlsx',
                            sheet_name='ant_1_3',
                            index_col=0)
```

This fragment of the code uses the `joblib` library to load an ML model. In this case, it is a model that we used in *Chapter 10* when we trained a classical ML model. It is a decision tree model.

Then, the program reads the same dataset that we used for training the model so that the format of the data is exactly the same. In this case, we can expect the same results that we used for the training dataset. For more complex models, we can create such a table by making one inference directly after the model has been trained and before it was saved.

Now, we can define three test cases in the following code fragment:

```
# test that the model is not null
# which means that it actually exists
def test_model_not_null():
    assert model is not None

# test that the model predicts class 1 correctly
# here correctly means that it predicts the same way as when it was
trained
def test_model_predicts_class_correctly():
    X = dfDataAnt13.drop(['Defect'], axis=1)
    assert model.predict(X)[0] == 1

# test that the model predicts class 0 correctly
# here correctly means that it predicts the same way as when it was
trained
def test_model_predicts_class_0_correctly():
    X = dfDataAnt13.drop(['Defect'], axis=1)
    assert model.predict(X)[1] == 0
```

The first test function (`test_model_not_null`) checks if the `model` variable, which is expected to hold the trained ML model, is not `null`. If the model is `null` (that is, it does not exist), the `assert` statement will raise an exception, indicating that the test has failed.

The second test function (`test_model_predicts_class_correctly`) checks whether the model predicts class 1 correctly for the given dataset. It does so by doing the following:

- Preparing the X input features by dropping the `'Defect'` column from the `dfDataAnt13` DataFrame, assuming that `'Defect'` is the target column (class label).

- Using the trained model (`model.predict(X)`) to make predictions on the X input features.

- Asserting that the first prediction (`model.predict(X)[0]`) should be equal to 1 (class 1). If the model predicts class 1 correctly, the test passes; otherwise, it raises an exception, indicating a test failure.

The third test case (`test_model_predicts_class_0_correctly`) checks whether the model predicts class 0 correctly for the given dataset. It follows a similar process as the previous test:

- Preparing the X input features by dropping the `'Defect'` column from the `dfDataAnt13` DataFrame.

- Using the trained model (`model.predict(X)`) to make predictions on the X input features.

- Asserting that the second prediction (`model.predict(X)[1]`) should be equal to 0 (class 0). If the model predicts class 0 correctly, the test passes; otherwise, it raises an exception, indicating a test failure.

These tests verify the integrity and correctness of the trained model and ensure it performs as expected on the given dataset. The output from executing the tests is shown as follows:

```
=============== test session starts ========================
platform win32 -- Python 3.11.4, pytest-7.4.0, pluggy-1.2.0
rootdir: C:\machine_learning_best_practices\chapter_12
plugins: anyio-3.7.0
collected 4 items

chapter_12_classical_ml_test.py                          [ 75%]
chapter_12_download_model_test.py                        [100%]


================= 4 passed in 12.76s ========================
```

The Pytest framework found all of our tests and showed that three (out of four) are in the `chapter_12_classical_ml_test.py` file and one is in the `chapter_12_downloaded_model_test.py` file.

My next best practice is, therefore, this:

> **Best practice #64**
> Treat models as units and prepare unit tests for them accordingly.

I recommend treating ML models as units (the same as modules) and using unit testing practices for them. This helps to reduce the effects of the probabilistic nature of the models and provides us with the possibility to check whether the model works correctly. It helps to debug the entire software system afterward.

Monitoring ML systems at runtime

Monitoring pipelines in production is a critical aspect of MLOps to ensure the performance, reliability, and accuracy of deployed ML models. This includes several practices.

The first practice is logging and collecting metrics. This activity includes instrumenting the ML code with logging statements to capture relevant information during model training and inference. Key metrics to monitor are model accuracy, data drift, latency, and throughput. Popular logging and monitoring frameworks include Prometheus, Grafana, and **Elasticsearch, Logstash, and Kibana (ELK)**.

The second one is alerting, which is a setup of alerts based on predefined thresholds for key metrics. This helps in proactively identifying issues or anomalies in the production pipeline. When an alert is triggered, the appropriate team members can be notified to investigate and address the problem promptly.

Data drift detection is the third activity, which includes monitoring the distribution of incoming data to identify data drift. Data drift refers to changes in data distribution over time, which can impact model performance.

The fourth activity is performance monitoring, where the MLOps team continuously tracks the performance of the deployed model. They measure inference times, prediction accuracy, and other relevant metrics, and they monitor for performance degradation, which might occur due to changes in data, infrastructure, or dependencies.

In addition to these four main activities, an MLOps team has also the following responsibilities:

- **Error analysis**: Using tools to analyze and log errors encountered during inference and understanding the nature of errors can help improve the model or identify issues in the data or system.

- **Model versioning**: Keep track of model versions and their performance over time, and (if needed) roll back to previous versions if issues arise with the latest deployment.

- **Environment monitoring**: Monitoring the infrastructure and environment where the model is deployed with KPIs such as CPU/memory utilization, and network traffic and looking for performance bottlenecks.

- **Security and compliance**: Ensuring that the deployed models adhere to security and compliance standards as well as monitor access logs and any suspicious activities.

- **User feedback**: Collecting, analyzing, and incorporating user feedback into the monitoring and inference process. MLOps solicits feedback from end users to understand the model's performance from a real-world perspective.

By monitoring pipelines effectively, MLOps can quickly respond to any issues that arise, deliver better user experiences, and maintain the overall health of your ML systems. However, monitoring all of the aforementioned aspects is rather effort-intensive, and not all MLOps teams have the resources to do that. Therefore, my last best practice in this chapter is this:

> **Best practice #65**
> Identify key aspects of the ML deployment and monitor these aspects accordingly.

Although this sounds straightforward, it is not always easy to identify key aspects. I usually start by prioritizing the monitoring of the infrastructure and logging and collecting metrics. Monitoring of the infrastructure is important as any kind of problems quickly propagate to customers and result in losing credibility and even business. Monitoring metrics and logging gives a great insight into the operation of ML systems and prevents a lot of problems with the production of ML systems.

Summary

Constructing ML pipelines concludes the part of the book that focuses on the core technical aspects of ML. Pipelines are important for ensuring that the ML models are used according to best practices in software engineering.

However, ML pipelines are still not a complete ML system. They can only provide inference of the data and provide an output. For the pipelines to function effectively, they need to be connected to other parts of the system such as the user interface and storage. That is the content of the next chapter.

References

- *A. Lima, L. Monteiro*, and *A.P. Furtado, MLOps: Practices, Maturity Models, Roles, Tools, and Challenges-A Systematic Literature Review. ICEIS (1), 2022: p. 308-320.*

- *John, M.M., Olsson, H.H.*, and *Bosch, J., Towards MLOps: A framework and maturity model.* In *2021 47th Euromicro Conference on Software Engineering and Advanced Applications (SEAA). 2021. IEEE.*

- *Staron, M. et al., Industrial experiences from evolving measurement systems into self-healing systems for improved availability. Software: Practice and Experience, 2018. 48(3): p. 719-739.*

13

Designing and Implementing Large-Scale, Robust ML Software

So far, we have learned how to develop ML models, how to work with data, and how to create and test the entire ML pipeline. What remains is to learn how we can integrate these elements into a **user interface** (**UI**) and how to deploy it so that they can be used without the need to program. To do so, we'll learn how to deploy the model complete with a UI and the data storage for the model.

In this chapter, we'll learn how to integrate the ML model with a graphical UI programmed in Gradio and storage in a database. We'll use two examples of ML pipelines – an example of the model for predicting defects from our previous chapters and a generative AI model to create pictures from a natural language prompt.

In this chapter, we're going to cover the following main topics:

- ML is not alone – elements of a deployed ML-based system

- The UI of an ML model

- Data storage

- Deploying an ML model for numerical data

- Deploying a generative ML model for images

- Deploying a code completion model as an extension to Visual Studio Code

ML is not alone

Chapter 2 introduced several elements of an ML system – storage, data collection, monitoring, and infrastructure, just to name a few of them. We need all of them to deploy a model for the users, but not all of them are important for the users directly. We need to remember that the users are interested in the results, but we need to pay attention to all details related to the development of such systems. These activities are often called AI engineering.

The UI is important as it provides the ability to access our models. Depending on the use of our software, the interface can be different. So far, we've focused on the models themselves and on the data that is used to train the models. We have not focused on the usability of models and how to integrate them into the tools.

By extension, as for the UI, we also need to talk about storing data in ML. We can use **comma-separated values (CSV)** files, but they quickly become difficult to handle. They are either too large to read into memory or too cumbersome for version control and exchanging data.

Therefore, in this chapter, we'll focus on making the ML system usable. We'll learn how to develop a UI, how to link the system to the database, and how to design a Visual Studio Code extension that can complete code in Python.

The UI of an ML model

A UI serves as the bridge between the intricate complexities of ML algorithms and the end users who interact with the system. It is the interactive canvas that allows users to input data, visualize results, control parameters, and gain insights from the ML model's outputs. A well-designed UI empowers users, regardless of their technical expertise, to harness the potential of ML for solving real-world problems.

Effective UIs for ML applications prioritize clarity, accessibility, and interactivity. Whether the application is aimed at business analysts, healthcare professionals, or researchers, the interface should be adaptable to the user's domain knowledge and objectives. Clear communication of the model's capabilities and limitations is vital, fostering trust in the technology and enabling users to make informed decisions based on its outputs. Hence my next best practice.

> **Best practice #66**
> Focus on the user task when designing the UI of the ML model.

We can use different types of UIs, but the majority of modern tools gravitate around two – web-based interfaces (which require thin clients) and extensions (which provide in-situ improvements). ChatGPT is an example of the web-based interface to the GPT-4 model, while GitHub CoPilot is an example of the extension interface to the same model.

In the first example, let's look at how easy it is to deploy an ML app using the Gradio framework. Once we have prepared a pipeline for our model, we just need a handful of lines of code to make the app. Here are the lines, based on the example of a model that exists at Hugging Face, for text classification:

```
import gradio as gr
from transformers import pipeline

pipe = pipeline("text-classification")

gr.Interface.from_pipeline(pipe).launch()
```

The first two lines import the necessary libraries – one for the UI (Gradio) and one for the pipeline. The second line imports the default text classification pipeline from Hugging Face and the last line creates the UI for the pipeline. The UI is in the form of a website with input and output buttons, as shown in *Figure 13.1*:

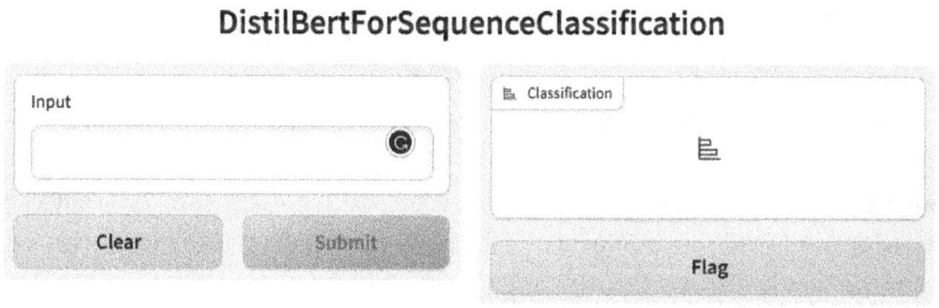

Figure 13.1 – UI for the default text classification pipeline

We can test it by inputting some example text. Normally, we would input this in a script and provide some sort of analysis, but this is done by the Gradio framework for us. We do not even need to link the parameters of the pipeline with the elements of the UI.

What happens behind the scenes can be explained by observing the output of the script in the console (edited for brevity):

```
No model was supplied, defaulted to distilbert-base-uncased-finetuned-
sst-2-english and revision af0f99b
Using a pipeline without specifying a model name and revision in
production is not recommended.
Downloading (...)lve/main/config.json: 100%|█████████████████| 629/629
[00:00<00:00, 64.6kB/s]
Downloading model.safetensors:
100%|█████████████████| 268M/268M [00:04<00:00, 58.3MB/s]
Downloading (...)okenizer_config.json: 100%|█████████████████| 48.0/48.0
[00:00<00:00, 20.7kB/s]
```

```
Downloading (...) solve/main/vocab.txt: 100%|          | 232k/232k
[00:00<00:00, 6.09MB/s]
Running on local URL:  http://127.0.0.1:7860
```

The framework has downloaded the default model, its tokenizers, and the vocabulary file and then created the application on the local machine.

The result of using this app is presented in *Figure 13.2*. We input some simple text and almost instantly get its classification:

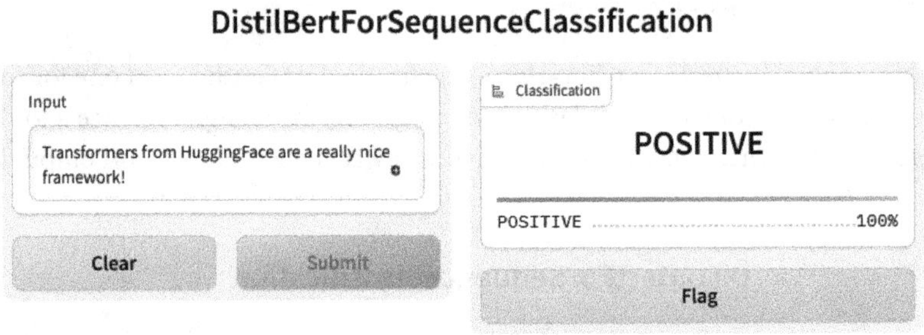

Figure 13.2 – Data analyzed using the default text classification pipeline

This kind of integration is a great way to deploy models first and to make sure that they can be used without the need to open a Python environment or similar. With this, we've come to my next best practice.

> **Best practice #67**
>
> Prepare your models for web deployment.

Regardless of what kind of models you develop, try to prepare them for web deployment. Our models can be then packaged as Docker containers and provided as part of a larger system of microservices. Using Gradio is a great example of how such a web deployment can be achieved.

Data storage

So far, we've used CSV files and Excel files to store our data. It's an easy way to work with ML, but it is also a local one. However, when we want to scale our application and use it outside of just our machine, it is often much more convenient to use a real database engine. The database plays a crucial role in an ML pipeline by providing a structured and organized repository for storing, managing, and retrieving data. As ML applications increasingly rely on large volumes of data, integrating a database into the pipeline becomes essential for a few reasons.

Databases offer a systematic way to store vast amounts of data, making it easily accessible and retrievable. Raw data, cleaned datasets, feature vectors, and other relevant information can be efficiently stored in the database, enabling seamless access by various components of the ML pipeline.

In many ML projects, data preprocessing is a critical step that involves cleaning, transforming, and aggregating data before feeding it to the model. Databases allow you to store intermediate preprocessed data, reducing the need to repeat resource-intensive preprocessing steps each time the model is trained. This speeds up the overall pipeline and maintains data consistency.

ML pipelines often involve data from diverse sources such as sensors, APIs, files, and databases. Having a centralized database simplifies the process of integrating different data streams, ensuring that all relevant information is readily available for training and inference.

Even maintaining a record of different dataset versions is important for reproducibility and tracking changes. Databases can be used to store different versions of datasets, making it easier to roll back to previous versions if needed and facilitating collaboration among team members.

Finally, ML applications that handle large-scale data require efficient data management to scale effectively. Databases provide mechanisms for indexing, partitioning, and optimizing queries, which enhance performance and allow the pipeline to handle increasing data volumes.

So, let's create a database in SQLite that would contain the same numerical data that we used in our previous work:

```
# create the database
import sqlite3


conn = sqlite3.connect('ant13.db')


c = conn.cursor()
```

In the preceding code fragment, we use the `sqlite3` engine to create a database and connect to it (`sqlite3.connect`). Once we connect to a database, we need a cursor to move around in the database and execute our queries. The next step is to import our existing data into the database.

Now, we can open the Excel file and transfer the data to the database:

```
# read the excel file with the data
# and save the data to the database
import pandas as pd


# read the excel file
df = pd.read_excel('chapter_12.xlsx', sheet_name='ant_1_3')


# print the first 5 rows
print(df.head())


# create the engine that we use to connect to the database to
# save the data
engine = create_engine('sqlite:///ant13.db')
```

```
# save the dataframe to the database
df.to_sql('ant_1_3', engine, index=False, if_exists='replace')
```

The preceding code reads data from an Excel file, processes it using the pandas library, and then saves the processed data into an SQLite database. First, the code reads an Excel file called `'chapter_12.xlsx'` and extracts data from the `'ant_1_3'` sheet. The data is loaded into a pandas DataFrame, df. Then, the code uses the `create_engine` function from the `sqlalchemy` library to establish a connection to an SQLite database. It then creates a connection to a database file named `'ant13.db'`.

Then, it uses the built-in `to_sql` function to create a database table based on the DataFrame. In this example, the function has the following parameters:

- `'ant_1_3'` is the name of the table in the database where the data will be stored.

- `engine` is the connection to the SQLite database that was created earlier.

- `index=False` specifies that the DataFrame index should not be saved as a separate column in the database.

- `if_exists='replace'` indicates that if a table named `'ant_1_3'` already exists in the database, it should be replaced with the new data. Other options for `if_exists` include `append` (add data to the table if it exists) and `fail` (raise an error if the table already exists).

After this, we have our data in a database and can easily share the data across multiple ML pipelines. However, in our case, we'll only demonstrate how to extract such data into a DataFrame so that we can use it in a simple ML application:

```
# select all rows from that database
data = engine.execute('SELECT * FROM ant_1_3').fetchall()
# and now, let's create a dataframe from that data
df = pd.DataFrame(data)

# get the names of the columns from the SQL database
# and use them as the column names for the dataframe
df.columns = [x[0] for x in engine.description]

# print the head of the dataframe
df.head()
```

The `'SELECT * FROM ant_1_3'` query selects all columns from the `'ant_1_3'` table in the database. The `fetchall()` method retrieves all the rows returned by the query and stores them in the data variable. The data variable will be a list of tuples, where each tuple represents a row of data.

Then, it creates a pandas DataFrame, df, from the data list. Each tuple in the list corresponds to a row in the DataFrame, and the columns of the DataFrame will be numbered automatically. Finally, the code retrieves the names of the columns in the original database table. The `engine.description` attribute holds metadata about the result of the executed SQL query. Specifically, it provides information about the columns returned by the query. The code then extracts the first element of each tuple in `engine.description`, which is the column name, and assigns these names to the columns of the DataFrame, `df`.

From there, the workflow with the data is just as we know it – it uses a pandas DataFrame.

In this example, the entire DataFrame fits in the database and the entire database can fit into one frame. However, this is not the case for most ML datasets. The pandas library has limitations in terms of its size, so when training models such as GPT models, we need more data than a DataFrame can hold. For that, we can use either the Dataset library from Hugging Face, or we can use databases. We can only fetch a limited amount of data, train a neural network on it, validate on another data, then fetch a new set of rows, train the neural network a bit more, and so on.

In addition to making the database on files, which can be a bit slow, the SQLite library allows us to create databases in memory, which is much faster, but they do not get serialized to our permanent storage – we need to take care of that ourselves.

To create an in-memory database, we can simply change the name of the database to `:memory:` in the first script, like this:

```
conn = sqlite3.connect(':memory:')

c = conn.cursor()
```

We can use it later on in a similar way, like so:

```
# create the enginve that we use to connect to the database to
# save the data
engine = create_engine('sqlite:///:memory:')

# save the dataframe to the database
df.to_sql('ant_1_3', engine, index=False, if_exists='replace')
```

In the end, we need to remember to serialize the database to a file; otherwise, it will disappear the moment our system closes:

```
# serialize to disk
c.execute("vacuum main into 'saved.db'")
```

Using databases together with ML is quite simple if we know how to work with DataFrames. The added value, however, is quite large. We can serialize data to files, read them into memory, manipulate them, and serialize them again. We can also scale up our applications beyond one system and use these systems online. However, for that, we need a UI.

With that, we've come to my next best practice.

> **Best practice #68**
>
> Try to work with in-memory databases and dump them to disk often.

Libraries such as pandas have limitations on how much data they can contain. Databases do not. Using an in-memory database provides a combination of the benefits of both without these limitations. Storing the data in memory enables fast access, and using the database engine does not limit the size of the data. We just need to remember to save (dump) the database from the memory to the disk once in a while to prevent the loss of data in case of exceptions, errors, defects, or equipment failures.

Deploying an ML model for numerical data

Before we create the UI, we need to define a function that will take care of making predictions using a model that we trained in the previous chapter. This function takes the parameters as a user would see them and then makes a prediction. The following code fragment contains this function:

```
import gradio as gr
import pandas as pd
import joblib

def predict_defects(cbo,
                    dcc,
                    exportCoupling,
                    importCoupling,
                    nom,
                    wmc):

    # we need to convert the input parameters to floats to use them in
the prediction
    cbo = float(cbo)
    dcc = float(dcc)
    exportCoupling = float(exportCoupling)
    importCoupling = float(importCoupling)
    nom = float(nom)
    wmc = float(wmc)

    # now, we need to make a data frame out of the input parameters
    # this is necessary because the model expects a data frame
    # we create a dictionary with the column names as keys
    # and the input parameters as values
    # please note that the names of the features must be the same as
in the model
```

```
data = {
    'CBO': [cbo],
    'DCC': [dcc],
    'ExportCoupling': [exportCoupling],
    'ImportCoupling': [importCoupling],
    'NOM': [nom],
    'WMC': [wmc]
}

# we create a data frame from the dictionary
df = pd.DataFrame(data)

# load the model
model = joblib.load('./chapter_12_decision_tree_model.joblib')

# predict the number of defects
result = model.predict(df)[0]

# return the number of defects
return result
```

This fragment starts by importing three libraries that are important for the UI and the modeling. We already know about the pandas library, but the other two are as follows:

- gradio: This library is used to create simple UIs for interactive ML model testing. The library makes it very easy to create the UI and connect it to the model.

- joblib: This library is used for saving and loading Python objects, particularly ML models. Thanks to this library, we do not need to train the model every time the user wants to open the software (UI).

The predict_defects function is where we use the model. It is important to note that the naming of the parameters is used automatically by the UI to name the input boxes (as we'll see a bit later). It takes six input parameters: cbo, dcc, exportCoupling, importCoupling, nom, and wmc. These parameters are the same software metrics that we used to train the model. As these parameters are input as text or numbers, it is important to convert them into floats, as this was the input value of our model. Once they have been converted, we need to turn these loose parameters into a single DataFrame that we can use as input to the model. First, we must convert it into a dictionary and then use that dictionary to create a DataFrame.

Once the data is ready, we can load the model using the model = joblib.load('./chapter_12_decision_tree_model.joblib') command. The last thing we must do is make a prediction using that model. We can do this by writing result = model.predict(df)[0]. The function ends by returning the result of the predictions.

There are a few items that are important to note. First, we need a separate function to handle the entire workflow since the UI is based on that. This function must have the same number of parameters as the number of input elements we have on our UI. Second, it is important to note that the names of the columns in the DataFrame should be the same as the names of the columns in the training data (the names are case-sensitive).

So, the actual UI is handled completely by the Gradio library. This is exemplified in the following code fragment:

```
# This is where we integrate the function above with the user
interface
# for this, we need to create an input box for each of the following
parameters:
# CBO, DCC, ExportCoupling,  ImportCoupling,  NOM,  WMC

demo = gr.Interface(fn=predict_defects,
                    inputs = ['number', 'number', 'number', 'number',
'number', 'number'],
                    outputs = gr.Textbox(label='Will contain
defects?',
                                         value= 'N/A'))

# and here we start the actual user interface
# in a browser window
demo.launch()
```

This code fragment demonstrates the integration of the previously defined `predict_defects` function with a UI. Gradio is used to create a simple UI that takes input from the user, processes it using the provided function, and displays the result. The code consists of two statements:

1. Creating the interface using the `gr.Interface` function with the following parameters:

    *   `fn=predict_defects`: This argument specifies the function that will be used to process the user input and produce the output. In this case, it's the `predict_defects` function that was defined previously. Please note that the arguments of the function are not provided, and the library takes care of extracting them (and their names) automatically.

    *   `inputs`: This argument specifies the types of inputs the interface should expect. In this case, it lists six input parameters, each of the `'number'` type. These correspond to the `cbo`, `dcc`, `exportCoupling`, `importCoupling`, `nom`, and `wmc` parameters in the `predict_defects` function.

    *   `outputs`: This argument specifies the output format that the interface should display to the user. In this case, it's a text box labeled **'Will contain defects?'** with an initial value of `'N/A'`. Since our model is binary, we only use 1 and 0 as the output. To mark the fact that the model has not been used yet, we start with the `'N/A'` label.

2. Launching the interface (`demo.launch()`): This line of code starts the UI in a web browser window, allowing users to interact with it.

The UI that was created using Gradio has input fields where the user can provide values for the software metrics (`cbo`, `dcc`, `exportCoupling`, `importCoupling`, `nom`, `wmc`). Once the user provides these values and submits the form, the `predict_defects` function will be called with the provided input values. The predicted result (whether defects will be present or not) will be displayed in the text box labeled 'Will contain defects?'

We can start this application by typing the following in the command prompt:

```
>python app.py
```

This starts a local web server and provides us with the address of it. Once we open the page with the app, we'll see the following UI:

Figure 13.3 – UI for the defect prediction model created using Gradio

The UI is structured into two columns – the right-hand column with the result and the left-hand column with the input data. At the moment, the input data is the default, and therefore the prediction value is N/A, as per our design.

We can fill in the data and press the **Submit** button to obtain the values of the prediction. This is shown in *Figure 13.4*:

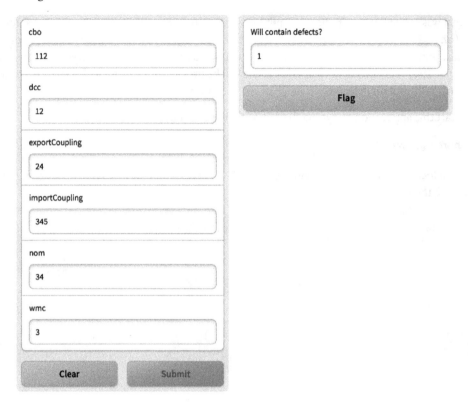

Figure 13.4 – UI with the prediction outcome

Once we fill in data to make predictions, we can submit it; at this point, our outcome shows that the module with these characteristics would contain defects. It's also quite logical – for any module that has 345 inputs, we could almost guarantee that there would be some defects. It's just too complex.

This model, and the UI, are only available locally on our computer. We can, however, share it with others and even embed it in websites, if we change just one line. Instead of `demo.launch()` without parameters, we can supply one parameter – `demo.launch(share=True)`.

Although we've used Gradio as an example of the UI, it illustrates that it is rather easy to link an existing model to a UI. We can input the data manually and get a prediction from the model. Whether the UI is programmed in Gradio or any other framework becomes less important. The difficulty may differ – for example, we may need to program the link between the input text boxes and model parameters manually – but the essence is the same.

Deploying a generative ML model for images

The Gradio framework is very flexible and allows for quickly deploying models such as generative AI stable diffusion models – image generators that work similarly to the DALL-E model. The deployment of such a model is very similar to the deployment of the numerical model we covered previously.

First, we need to create a function that will generate images based on one of the models from Hugging Face. The following code fragment shows this function:

```
import gradio as gr
import pandas as pd
from diffusers import StableDiffusionPipeline
import torch

def generate_images(prompt):
    '''
    This function uses the prompt to generate an image
    using the anything 4.0 model from Hugging Face
    '''

    # importing the model from Hugging Face
    model_id = "xyn-ai/anything-v4.0"
    pipe = StableDiffusionPipeline.from_pretrained(model_id,
                                        torch_dtype=torch.
float16,

                                        safety_
checker=None)

    # send the pipeline to the GPU for faster processing
    pipe = pipe.to("cuda")

    # create the image here
    image = pipe(prompt).images[0]

    # return the number of defects
    return image
```

This code fragment starts by importing the necessary libraries. Here, we'll notice that there is another library – diffusers – which is an interface to image generation networks. The function imports a pre-trained model from the Hugging Face hub. The model is "xyn-ai/anything-v4.0". It is a variant of the Anything 4.0 model, cloned by one of the users. The StableDiffusionPipeline. from_pretrained() function is used to load the model as a pipeline for image generation. The torch_dtype parameter is set to torch.float16, which indicates the data type to be used for computations (lower precision for faster processing).

The image is generated using the pipeline bypassing the prompt as an argument to the `pipe()` function. The generated images are accessed using the `images[0]` attribute. The `prompt` parameter is provided through the parameter of the function, which is supplied by the UI.

The function returns the image, which is then captured by the UI and displayed.

The code for the UI is also quite straightforward once we know the code from the previous example:

```
demo = gr.Interface(fn=generate_images,
                    inputs = 'text',
                    outputs = 'image')


# and here we start the actual user interface
# in a browser window
demo.launch()
```

Compared to the previous example, this code contains only one input parameter, which is the prompt that's used to generate the image. It also has one output, which is the image itself. We use the `'image'` class to indicate that it is an image and should be displayed as such. The output of this model is presented in *Figure 13.5*:

Figure 13.5 – Image generated from the Anything 4.0 model using the "car and sun" prompt

Please note that the model is not perfect as the generated car has distortion artifacts – for example, the right-hand taillight has not been generated perfectly.

Deploying a code completion model as an extension

So far, we've learned how to deploy models online and on the Hugging Face hub. These are good methods and provide us with the ability to create a UI for our models. However, these are standalone tools that require manual input and provide an output that we need to use manually – for example, paste into another tool or save to disk.

In software engineering, many tasks are automated and many modern tools provide an ecosystem of extensions and add-ins. GitHub Copilot is such an add-in to Visual Studio 2022 and an extension to Visual Studio Code – among other tools. ChatGPT is both a standalone web tool and an add-in to Microsoft's Bing search engine.

Therefore, in the last part of this chapter, we'll package our models as an extension to a programming environment. In this section, we learn how to create an extension to complete code, just like GitHub CoPilot. Naturally, we won't use the CodeX model from CoPilot, but Codeparrot's model for the Python programming language. We've seen this model before, so let's dive deeper into the actual extension.

We need a few tools to develop the extension. Naturally, we need Visual Studio Code itself and the Python programming environment. We also need the Node.js toolkit to create the extension. We installed it from nodejs.org. Once we have installed it, we can use Node.js's package manager to install Yeoman and the framework to develop the extension. We can do that by using the following command in the command prompt:

```
npm install -g yo generator-code
```

Once the packages have been installed, we need to create the skeleton code for our extension by typing the following:

```
yo code
```

This will bring up the menu that we need to fill in:

```
    _-----_
   |       |    ╭──────────────────────────╮
   |--(o)--|    |   Welcome to the Visual  |
  `---------´   |   Studio Code Extension  |
   ( _´U`_ )    |        generator!        |
   /___A___\   /╰──────────────────────────╯
    |  ~  |
  __´.___.´__
 ´   `  |° ´ Y `

? What type of extension do you want to create? (Use arrow keys)
> New Extension (TypeScript)
  New Extension (JavaScript)
```

```
New Color Theme
New Language Support
New Code Snippets
New Keymap
New Extension Pack
New Language Pack (Localization)
New Web Extension (TypeScript)
New Notebook Renderer (TypeScript)
```

We need to choose the first option, which is a new extension that uses Typescript. It is the easiest way to start writing the extension. We could develop a very powerful extension using the language pack and language protocol, but for this first extension, simplicity beats power.

We need to make a few decisions about the setup of our extension, so let's do that now:

```
? What type of extension do you want to create? New Extension
(TypeScript)
? What's the name of your extension? mscopilot
? What's the identifier of your extension? mscopilot
? What's the description of your extension? Code generation using
Parrot
? Initialize a git repository? No
? Bundle the source code with webpack? No
? Which package manager to use? (Use arrow keys)
> npm
  yarn
  pnpm
```

We call our extension `mscopilot` and do not create much additional code – no Git repository and no webpack. Again, simplicity is the key for this example. Once the folder has been created, we need one more package from Node.js to interact with Python:

```
npm install python-shell
```

After we click on the last entry, we get a new folder named `mscopilot`; we can enter it with the `code .` command. It opens Visual Studio Code, where we can fill the template with the code for our new extension. Once the environment opens, we need to navigate to the `package.json` file and change a few things. In that file, we need to find the `contributes` section and make a few changes, as shown here:

```
"contributes": {
    "commands": [
        {
            "command": "mscopilot.logSelectedText",
            "title": "MS Suggest code"
```

```
        }
    ],
    "keybindings": [
        {
            "command": "mscopilot.logSelectedText",
            "key": "ctrl+shift+l",
            "mac": "cmd+shift+l"
        }
    ]
},
```

In the preceding code fragment, we added some information stating that our extension has one new function – logSelectedText – and that it will be available via the *Ctrl + Shift + L* key combination on Windows (and a similar one on Mac). We need to remember that the command name includes the name of our extension so that the extension manager knows that this command belongs to our extension. Now, we need to go to the extension.ts file and add the code for our command. The code following fragment contains the first part of the code – the setup for the extension and its activation:

```
import * as vscode from 'vscode';

// This method is called when your extension is activated
export function activate(context: vscode.ExtensionContext) {
    // Use the console to output diagnostic information (console.log)
    and errors (console.error)
    // This line of code will only be executed once when your extension
    is activated
    console.log('Congratulations, your extension "mscopilot" is now
    active!');
```

This function just logs that our extension has been activated. Since the extension is rather invisible to the user (and it should be), it is a good practice to use the log file to store the information that has been instantiated.

Now, we add the code that will get the selected text, instantiate the Parrot model, and add the suggestion to the editor:

```
// Define a command to check which code is selected.
vscode.commands.registerCommand('mscopilot.logSelectedText', () => {
    // libraries needed to execute python scripts
    const python = require('python-shell');
    const path = require('path');

    // set up the path to the right python interpreter
    // in case we have a virtual environment
    python.PythonShell.defaultOptions = { pythonPath: 'C:/Python311/
```

```
python.exe' };
  // Get the active text editor
  const editor = vscode.window.activeTextEditor;

  // Get the selected text
     const selectedText = editor.document.getText(editor.selection);

  // prompt is the same as the selected text
  let prompt:string = selectedText;

  // this is the script in Python that we execute to
  // get the code generated by the Parrot model
  //
  // please note the strange formatting,
  // which is necessary as python is sensitive to indentation
  let scriptText = `
from transformers import pipeline

pipe = pipeline("text-generation", model="codeparrot/codeparrot-
small")
outputs = pipe("${prompt}", max_new_tokens=30, do_sample=False)
print(outputs[0]['generated_text'])`;

  // Let the user know what we start the code generation
  vscode.window.showInformationMessage(`Starting code generation for
prompt: ${prompt}`);

  // run the script and get the message back
  python.PythonShell.runString(scriptText, null).then(messages=>{
  console.log(messages);

  // get the active editor to paste the code there
  let activeEditor = vscode.window.activeTextEditor;

  // paste the generated code snippet
  activeEditor.edit((selectedText) => {

  // when we get the response, we need to format it
  // as one string, not an array of strings
  let snippet = messages.join('\n');

  // and replace the selected text with the output
  selectedText.replace(activeEditor.selection, snippet)  });
  }).then(()=>{
```

```
    vscode.window.showInformationMessage(`Code generation
finished!`);});
    });
context.subscriptions.push(disposable);
}
```

This code registers our `'mscopilot.logSelectedText'` command. We made this visible to the extension manager in the previous file – `package.json`. When this command is executed, it performs the following steps. The important part is the interaction between the code in TypeScript and the code in Python. Since we're using the Hugging Face model, the easiest way is to use the same scripts that we've used so far in this book. However, since the extensions are written in TypeScript (or JavaScript), we need to embed the Python code in TypeScript, add a variable to it, and capture the outcome:

1. First, imports the required libraries – `python-shell` and `path` – which are needed to execute Python scripts from within a Node.js environment.

2. Next, it sets up the Python interpreter via `C:/Python311/python.exe`, which will be used to run Python scripts. This is important for ensuring that the correct Python environment is used, even when using a virtual environment. If we do not specify it, we need to find it in the script, which is a bit tricky in the user environment.

3. After, it sets the active text editor and the selected text. We need this selection so that we can send the prompt to the model. In our case, we'll simply send the selection to the model and get the suggested code.

4. Then, it prepares the prompt, which means that it creates a string variable that we use in the code of the Python script.

5. Next, it defines the Python script, where the connection to the ML model is established. Our Python script is defined as a multi-line string (`scriptText`) using a template literal. This script utilizes the Hugging Face Transformers library's `pipeline` function to perform text generation using the `codeparrot-small` model. The Python code is in boldface, and we can see that the string is complemented with the prompt, which is the selected text in the active editor.

6. Then, it displays short information to the user since the model requires some time to load and make an inference. It may take up to a minute (for the first execution) to get the inference as the model needs to be downloaded and set up. Therefore, it is important to display a message that we're starting the inference. A message is displayed to the user using `vscode.window.showInformationMessage`, indicating that the code generation process is about to start.

7. After, it runs the Python script (`scriptText`) using `python.PythonShell.runString`. The script's output is captured in the `messages` array. We lost control over the execution for a while since we waited for the Python script to finish; it provided us with a suggestion for code completion.

8. Next, it pastes the generated code from the response (`messages`) array into a single string (`snippet`). The snippet is then pasted into the active text editor at the position of the selected text, effectively replacing the selected text with the generated code. Since the first element of the response from the model is the prompt, we can simply just replace the selection with the snippet.

9. Finally, it displays the completion message after the code generation process.

Now, we have an extension that we can test. We can execute it by pressing the *F5* key. This brings up a new instance of Visual Studio, where we can type a piece of code to be completed, as shown in *Figure 13.6*:

Figure 13.6 – A test instance of some Visual Studio code with our extension
activated. The selected text is used as the prompt for the model

Once we press *Ctrl + Shift + L*, as we defined in the `package.json` file, our command is activated. This is indicated by the messages in the lower right-hand corner of the environment in *Figure 13.7*:

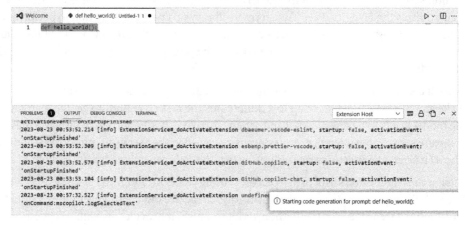

Figure 13.7 – Starting to generate the code. The message box and the
log information indicate that our command is working

After a few seconds, we get a suggestion from the Parrot model, which we must then paste into the editor, as shown in *Figure 13.8*:

Figure 13.8 – Code suggestion from the Parrot model pasted into the editor

Here, we can see that our extension is rather simple and gets all the suggestions from the model. So, in addition to the proper code (`return "Hello World!"`), it included more than we needed. We could write more interesting logic, parse the code, and clean it – the sky is the limit. I leave it to you to continue this work and to make it better. My job was to illustrate that writing a GitHub CoPilot-like tool is not as difficult as it may seem.

With this, we've come to my final best practice for this chapter.

> **Best practice #69**
>
> If your model/software aims to help in the daily tasks of your users, make sure that you develop it as an add-in.

Although we could use the Codeparrot model from the Gradio interface, it would not be appreciated. Programmers would have to copy their code to a web browser, click a button, wait for the suggestion, and paste it back into their environment. By providing an extension to Visual Studio Code, we can tap into the workflow of software developers. The only extra task is to select the text to complete and press *Ctrl + Shift + L*; I'm sure that this could be simplified even more, just like GitHub Copilot does.

Summary

This chapter concludes the third part of this book. It also concludes the most technical part of our journey through the best practices. We've learned how to develop ML systems and how to deploy them. These activities are often called AI engineering, which is the term that places the focus on the development of software systems rather than the models themselves. This term also indicates that testing, deploying, and using ML is much more than training, validating, and testing the models.

Naturally, there is even more to this. Just developing and deploying AI software is not enough. We, as software engineers or AI engineers, need to consider the implications of our actions. Therefore, in the next part of this book, we'll explore the concepts of bias, ethics, and the sustainable use of the fruits of our work – AI software systems.

References

- *Rana, R., et al. A framework for adoption of machine learning in industry for software defect prediction. In 2014 9th International Conference on Software Engineering and Applications (ICSOFT-EA). 2014. IEEE.*

- *Bosch, J., H.H. Olsson, and I. Crnkovic, Engineering ai systems: A research agenda. Artificial Intelligence Paradigms for Smart Cyber-Physical Systems, 2021: p. 1-19.*

- *Giray, G., A software engineering perspective on engineering machine learning systems: State of the art and challenges. Journal of Systems and Software, 2021. 180: p. 111031.*

Part 4:
Ethical Aspects of
Data Management and
ML System Development

Since machine learning is based on patterns in the training data, we need to understand how to work with this kind of system from an ethical perspective. Ethics is one of these areas that software engineers are not the most familiar with. In this part of the book, we look into what kind of ethical issues exist in data acquisition and management and how to work with bias in machine learning algorithms. Finally, we finish up this part by exploring how to integrate what we've learned in this book into an entire ecosystem of web services and how to deploy them.

This part has the following chapters:

- *Chapter 14, Ethics in Data Acquisition and Management*
- *Chapter 15, Ethics in Machine Learning Systems*
- *Chapter 16, Integration of ML Systems in Ecosystems*
- *Chapter 17, Summary and Where to Go Next*

14

Ethics in Data Acquisition and Management

Machine learning (ML) requires a lot of data that can come from a variety of sources, but not all sources are equally easy to use. In software engineering, we can design and develop systems that use data from other systems. We can also use data that does not really originate from people; for example, we can use data about defects or complexity of systems. However, to provide more value to society, we need to use data that contains information about people or their belongings; for example, when we train machines to recognize faces or license plates. Regardless of our use case, however, we need to follow ethical guidelines and, above all, have the guiding principle that our software should not cause any harm.

We start this chapter by exploring a few examples of unethical systems that show bias; for example, credit ranking systems that penalize certain minorities. I will also explain problems with using open source data and revealing the identities of subjects. The core of the chapter, however, is an explanation and discussion on ethical frameworks for data management and software systems, including the **Institute of Electrical and Electronics Engineers** (**IEEE**) and the **Association for Computing Machinery** (**ACM**) codes of conduct.

In this chapter, we're going to cover the following main topics:

- Ethics in computer science and software engineering
- Data is all around us, but can we really use it?
- Ethics behind data from open source systems
- Ethics behind data collected from humans
- Contracts and legal obligations

Ethics in computer science and software engineering

The modern view on ethics has its roots in the Nuremberg Code, which was developed after the Second World War. The code is based on several principles, but the most important one is the fact that every study needs to have permission if it involves human subjects. This is essential, as it prevents the abuse of humans during experimentation. Every participant in a study should also be able to retract their permission at any given time. Let us look at all 10 principles:

1. The voluntary consent of the human subject is absolutely essential.

2. The experiment should be such as to yield fruitful results for the good of society, unprocurable by other methods or means of study, and not random and unnecessary in nature.

3. The experiment should be so designed and based on the results of animal experimentation and a knowledge of the natural history of the disease or other problem under study that the anticipated results will justify the performance of the experiment.

4. The experiment should be so conducted as to avoid all unnecessary physical and mental suffering and injury.

5. No experiment should be conducted where there is an a priori reason to believe that death or disabling injury will occur except, perhaps, in those experiments where the experimental physicians also serve as subjects.

6. The degree of risk to be taken should never exceed that determined by the humanitarian importance of the problem to be solved by the experiment.

7. Proper preparations should be made, and adequate facilities provided to protect the experimental subject against even remote possibilities of injury, disability, or death.

8. The experiment should be conducted only by scientifically qualified persons. The highest degree of skill and care should be required through all stages of the experiment of those who conduct or engage in the experiment.

9. During the course of the experiment, the human subject should be at liberty to bring the experiment to an end if they have reached a physical or mental state where continuation of the experiment seems to them to be impossible.

10. During the course of the experiment, the scientist in charge must be prepared to terminate the experiment at any stage if they have probable cause to believe, in the exercise of the good faith, superior skill, and careful judgment required of them, that a continuation of the experiment is likely to result in injury, disability, or death of the experimental subject.

The Nuremberg Code laid the foundation for modern ethical standards in human experimentation and research. It has been influential in the development of subsequent ethical guidelines and regulations, such as the **Declaration of Helsinki (DoH)** and various national and international regulations governing human research ethics. These principles emphasize the importance of respecting the rights and well-being of research participants and ensuring that research is conducted in an ethical and responsible manner.

The aforementioned principles are designed to guide experiments toward bringing value to society, but at the same time to respect the participants of the experiments. The first principle is about consent, which is important as we want to prevent using subjects who are not aware of the experiments. In the context of ML, this means that we need to be very careful when we collect data that includes humans; for example, when we collect image data to train object recognition, or when we collect data from open source repositories that contain personal information.

Although we do conduct experiments in software engineering, these principles are more universal than we might think. In this chapter, we explore how these principles affect the ethics of data in the engineering of AI systems.

In addition to these principles, we also look at the sources of bias in data and how to avoid it.

Data is all around us, but can we really use it?

One of the ways in which we protect subjects and data is to use appropriate licenses for the use of data. Licenses are a sort of contract in that the licensor grants permission to use the data in a specific way to a licensee. Licenses are used for both software products (algorithms, components) and data. The following license models are the most commonly used ones in contemporary software:

- **Proprietary license**: It is a model whereby the licensor owns the data and grants permission to use the data for certain purposes, often for profit. In such a contract, the parties usually regulate what the data can be used for, how, and for how long. These licenses also regulate liabilities for both parties.

- **Permissive open licenses**: These are licenses that provide the licensee almost unrestricted access to the data, at the same time limiting the liability of the licensor. The licensee is often not required to provide access to the licensee's product or derivative work.

- **Non-permissive open licenses**: These are licenses that provide almost unrestricted access, at the same time requiring some sort of reciprocity. Usually, the reciprocity is in the form of requiring the licensee to provide access to the product or the derivative work.

Naturally, these three classes of licenses also have variants. So, let's take a look at one of the popular open source licenses – the Unsplash license from Hugging Face:

```
Unsplash
All unsplash.com images have the Unsplash license copied below:
https://unsplash.com/license
License
Unsplash photos are made to be used freely. Our license reflects that.

All photos can be downloaded and used for free
Commercial and non-commercial purposes
No permission needed (though attribution is appreciated!)
```

```
What is not permitted
Photos cannot be sold without significant modification.
Compiling photos from Unsplash to replicate a similar or competing
service.
Tip: How to give attribution
Even though attribution isn't required, Unsplash photographers
appreciate it as it provides exposure to their work and encourages
them to continue sharing.


Photo by <person name> on Unsplash


Longform
Unsplash grants you an irrevocable, nonexclusive, worldwide copyright
license to download, copy, modify, distribute, perform, and use photos
from Unsplash for free, including for commercial purposes, without
permission from or attributing the photographer or Unsplash. This
license does not include the right to compile photos from Unsplash to
replicate a similar or competing service.


Other Images
All other images were either taken by the authors, or created by
friends of the authors and all permissions to modify, distribute,
copy, perform and use are given to the authors.
```

The license comes from this dataset: `https://huggingface.co/datasets/google/dreambooth/blob/main/dataset/references_and_licenses.txt`. Let us take a closer look at what the license means. First of all, here's what the Unsplash license allows:

- **Free use**: We can download and use the images from Unsplash without any cost.

- **Commercial and non-commercial use**: We can use the images for both business-related purposes (such as in advertisements, websites, and product packaging) as well as personal projects or non-profit activities.

- **No need for permission**: We don't have to ask for consent or permission from the photographer or Unsplash to use the image. This makes the process hassle-free and convenient.

- **Modification and redistribution**: We can modify the original image in any way you see fit and distribute it. However, if you wish to sell the image, it should be significantly modified from the original.

- **Irrevocable, nonexclusive, worldwide license**: Once you download the photo, you have the right to use it indefinitely (irrevocable). Nonexclusive means others can also use the same image, and worldwide implies there are no geographical restrictions on its use.

At the same time, the license prohibits certain activities:

- **Selling unaltered photos**: We cannot sell the photos in their original form or without making significant modifications to them.

- **Replicating Unsplash's service**: We can't use the images to create a service that directly competes with or is similar to Unsplash. In other words, downloading a bunch of Unsplash images and then starting your own stock photo service using those images would be a violation of the license.

The license also regulates the form of attribution – while it's not mandatory to give credit to the photographer or Unsplash when you use an image, it is encouraged. Attribution is a way of recognizing the effort of the photographer. The provided example, "*Photo by <person name> on Unsplash*" is a suggested format for giving credit.

It also regulates what happens if there are other images in that dataset (for example, added by a third party): for images that are not from Unsplash, the authors either took these pictures themselves or had them created by acquaintances. They have full permission to use, distribute, modify, and perform these images without any restrictions.

We can also take a look at one of the permissive licenses from **Creative Commons** (**CC**); for example, the **Creative Commons, By Attribution, version 4.0** (**CC-BY-4.0**) license (`https://creativecommons.org/licenses/by/4.0/legalcode`). In short, this license allows the sharing and redistributing of data, under the following conditions:

- **Attribution**, which means that we must give the appropriate credit to the author of the data when we use it, and we must provide a link to the reference and indicate changes that we make to the data.

- **No additional restrictions**, which means that we cannot impose any additional restrictions on the use of the data that we license in this way. We must not use any means to restrict others from using this data if it is used in our product.

The entire text of the license is a bit too large to quote, but let us analyze parts of it so that we can see the difference from the very permissive Unsplash license. First of all, the license provides a number of definitions, which can help in legal disputes. For instance, *Section 1.i* defines what it means to share:

```
Share means to provide material to the public by any means or
process that requires permission under the Licensed Rights, such
as reproduction, public display, public performance, distribution,
dissemination, communication, or importation, and to make material
available to the public including in ways that members of the public
may access the material from a place and at a time individually chosen
by them.
```

The preceding legal text may sound complex, but when we read it, the text specifies what it means to share data. For example, it says that "*importation*" is one of the means of sharing data. This means that providing data as part of a Python package, C# library, or GitHub repository is a means of sharing data.

We can also take a look at *Section 2* of the license, where the definitions from *Section 1* are used. Here is an example of *Section 2.a.1*:

```
Subject to the terms and conditions of this Public License,
the Licensor hereby grants You a worldwide, royalty-free, non-
sublicensable, non-exclusive, irrevocable license to exercise the
Licensed Rights in the Licensed Material to:
* reproduce and Share the Licensed Material, in whole or in part; and
* produce, reproduce, and Share Adapted Material.
```

In the first part, we learn that the license allows us to use the material royalty-free; that is, we do not have to pay for it. It also specifies the geographical location for the use (worldwide) and says that we are not the only ones using this data (non-exclusive). Then, the license also says that we can reproduce the material, both in whole and in parts, or produce adapted material.

However, certain rights are not transferred to us, as licensees, which we learn in *Section 2.b.1*:

```
Moral rights, such as the right of integrity, are not licensed under
this Public License, nor are publicity, privacy, and/or other similar
personality rights; however, to the extent possible, the Licensor
waives and/or agrees not to assert any such rights held by the
Licensor to the limited extent necessary to allow You to exercise the
Licensed Rights, but not otherwise.
```

This section explains that moral rights to the data are not transferred. In particular, if we encounter data that is private or can be linked to persons, we are not granted the right to it. Three parts are important here:

- **Moral rights**: Moral rights are a subset of copyright that pertains to the personal and reputational, rather than the purely economic, aspects of a work. These rights might vary from one jurisdiction to another, but often they include the following:

  - **Right of integrity**: This is the right of an author to object to any distortion, mutilation, or other modification of a work if it would be prejudicial to their honor or reputation

  - **Right of attribution**: This is the right of an author to be credited as the creator of their work

  The CC license specifies that these moral rights are not licensed. This means that while someone might be able to use the work in ways defined by the CC license, they don't have carte blanche to modify the work in ways that could damage the original creator's reputation or honor.

- **Publicity, privacy, and other similar personality rights**: These rights pertain to an individual's personal data, likeness, name, or voice. They're rights that protect against unwanted exposure or exploitation. The CC license doesn't grant users the right to infringe on these either.

  For example, if a photograph of a person is under a CC license, while we might be able to use the photo itself in ways allowed by that license, it doesn't mean we can use the person's likeness in a commercial or promotional manner without their consent.

- **Waiving or not asserting rights**: The licensor is, however, waiving or agreeing not to enforce these moral or personal rights to the extent necessary for us to exercise the licensed rights. This means the licensor won't interfere with our use of the work as permitted by the CC license by invoking their moral or personal rights, but only up to a point. They're not giving up these rights entirely; they're only limiting their enforcement in contexts relevant to the CC license.

I strongly encourage readers to read the license and reflect on how it is constructed, including the parts that regulate the liability of the licensor.

We move on, however, to licenses that are not as permissive as the CC-BY-4.0 license. For example, let us take a look at one of the so-called copyleft licenses: **Attribution-NonCommercial-NoDerivatives 4.0 International** (**CC BY-NC-ND 4.0**). This license allows us to copy and redistribute data in any format, under the following conditions (quoted and adapted from the following CC web page: `https://creativecommons.org/licenses/by-nc-nd/4.0/`):

- **Attribution**: We must give appropriate credit, provide a link to the license, and indicate if changes were made. We may do so in any reasonable manner, but not in any way that suggests the licensor endorses us or our use.

- **NonCommercial**: We may not use the material for commercial purposes.

- **NoDerivatives**: If we remix, transform, or build upon the material, we may not distribute the modified material.

- **No additional restrictions**: We may not apply legal terms or technological measures that legally restrict others from doing anything the license permits.

The body of the license text is structured similarly to the permissive CC-BY-4.0 license, which we quoted before. Let us look at the part where the license differs – *Section 2.a.1*:

```
Subject to the terms and conditions of this Public License,
the Licensor hereby grants You a worldwide, royalty-free, non-
sublicensable, non-exclusive, irrevocable license to exercise the
Licensed Rights in the Licensed Material to:
* reproduce and Share the Licensed Material, in whole or in part, for
NonCommercial purposes only; and
* produce and reproduce, but not Share, Adapted Material for
NonCommercial purposes only.
```

The first bullet point means that we can share licensed material for non-commercial purposes only. The second bullet extends that to the adapted material as well.

One question that can be raised directly is, then, this: *How to select a license for the data that we create?* That's where my next best practice comes into play.

> **Best practice #70**
>
> If you create your own data for non-commercial purposes, use one of the permissive derivative licenses that limit your liability.

When designing commercial licenses, please always consult your legal counsel so that you select the best license model for your data and your product. However, if you work on open source data, try to use a license that regulates two aspects – whether it is possible to use your data for commercial purposes and what your liability is.

In my work, I try to be as open as possible, simply because of my belief in open science, but this does not need to be universally true. Therefore, I often guard myself from commercial use of my data – hence the non-commercial purposes.

The second aspect is liability. When we collect our data, we cannot think about all possible uses of data, which means that there is always a risk of the data being misused, or a risk of making a mistake when using our data. We do not want to be sued for something that we did not do, so it is always a good idea to limit our liability in such a license. We can limit it in a few ways; one of them is stating that the licensee is responsible for ensuring that the data is used in an ethical way or according to all rules and regulations applicable in the appropriate region.

This brings us to the next point – the ethical use of data from open source systems.

Ethics behind data from open source systems

Proprietary systems oftentimes have licenses that regulate who owns the data and for what purpose. For example, code review data from a company often belongs to the company. By working for the company, the employees usually sign off their rights to the data that they generate for the company. It is needed in the legal sense because the employees are getting compensated for that – usually in the form of salaries.

However, what the employees do not transfer to the company is the right to use their personal data freely. This means that when we work with source systems, such as the Gerrit review system, we should not extract personal information without the permission of the people involved. If we execute the query where masking of this data is not possible, we must ensure that the personal data is anonymized (as soon as it is possible) and is not leaked to the analysis. We must ensure that such personal data is not made publicly available.

One of the areas where we can find guidance is the field of mining software repositories; for example, in a recent article by Gold and Krinke. Despite a low sample size in this study, the authors touch upon important issues related to the ethical use of data from software repositories such as GitHub or Gerrit. They bring up several data sources, with the most popular being:

- **Version control data, such as CVS, Subversion, Git, or Mercurial**: There are challenges related to licensing and the presence of personal information in the repositories.

- **Issue tracker data, such as Bugzilla**: Challenges are mostly related to the presence of personal data in such repositories or the ability to connect data to persons implicitly.

- **Mail archives**: Mailing lists are flexible and can be used for different purposes; for example, issue tracking, code review, Q&A forums, and so on. However, they can contain sensitive personal information or can lead to conclusions that can impact individuals.

- **Build logs**: **Version control systems** (**VCS**) often use some kind of **continuous integration** (**CI**) system to automate building the software. If the build results are archived, they can provide data for research into testing and building practices.

- **Stack Overflow**: Stack Overflow provides official dumps of their data, and (subsets of) the dumps have been used as challenges directly, or inside a dataset aggregating historic information. Despite licensing regulations that require users to allow the use of their data for analyses, not all are ethical.

- **IDE events**: The main challenges are related to the fact that each IDE is set up for individuals and has access to very personal data and the behavior of the user.

Given the amount of source code that is available online in repositories such as GitHub, it is tempting to analyze it for all kinds of purposes. We could mine the repositories for source code in order to understand how software is constructed, designed, and tested. Such use of source code is then limited by the licenses provided for each repository.

At the same time, when mining data about source code, we can mine data about contributors, their comments for pull requests, and the organizations where they work. This use of repositories is then governed by the licenses and by the ethical guidelines for using personal data. As mentioned at the beginning of this chapter, using personal data requires consent from the individuals, which means that we should ask the persons whose data we analyze for permission. In practice, however, this is impossible, for two reasons – one is the practical ability to contact all contributors, and the second is the terms of use of the repository. Most repositories prohibit contacting contributors to study them.

Therefore, we need to apply the principle related to balance between privacy and the benefit. In most cases, the privacy of individuals is worth more than the value of the study that we conduct, hence my next best practice.

Best practice #71

Limit yourself to studying source code and other artifacts, and use personal data only with the consent of the subjects.

There are numerous ethical codes, but most of them come to the same conclusion – consent is required when using personal data. Therefore, we should use open source data with caution and analyze only the data that is necessary for our studies. Always obtain informed consent when analyzing data from people. As the article suggests, we can follow the next guidelines for studying repositories.

There are a number of guidelines for studying open source repositories:

- When studying source code, we need to be mostly concerned about the license and ensure that we do not mix source code with the personal data of the people.

- When studying commits, we need to respect the identities of people who make contributions, as they are probably not aware of being studied. We must make sure that no harm or damage is caused to these individuals based on our research. If we need to study personal data, then we need to apply for approval from the institutional ethics board and obtain explicit consent. In the case of some licenses, the license is extended to messages in commits – for example, Apache License 2.0: "*any form of electronic, verbal, or written communication sent to the Licensor or its representatives, including but not limited to communication on electronic mailing lists, source code control systems, and issue tracking systems*".

- When mining IDE events, we should obtain consent as the individuals are probably not aware of being studied. As opposed to the studies of data from open repositories, IDE events can contain more personal data, as these tools are integrated with the software ecosystem of the users.

- **When mining build logs**: The same principles apply here as they do for commits, as these can be easily linked.

- **When mining Stack Overflow**: Although the Stack Overflow license allows us to conduct certain research since the users need to allow this (according to the terms of use), we need to make sure that there is a balance between the benefits of the study and the risks related to analyzing the data, which is often a free text plus some source code.

- **When mining issue trackers**: Some issue trackers, such as Debian, provide the possibility to analyze the data, but it needs to be synchronized with the ethics board.

- When mining mailing lists, we need to try to obtain permission from the ethical board since these lists often contain personal information and should be scrutinized before it is used.

The Menlo Report by Kenneally and Dittrich from 2012 provides a number of guidelines for ethical research in information and communication technology, including software engineering. They define four principles (quoted after the report):

- **Respect for persons**: Participation as a research subject is voluntary, and follows from informed consent; Treat individuals as autonomous agents and respect their right to determine their own best interests; Respect individuals who are not targets of research yet are impacted; Individuals with diminished autonomy, who are incapable of deciding for themselves, are entitled to protection.

- **Beneficence**: Do no harm; Maximize probable benefits and minimize probable harm; systematically assess both risk of harm and benefit.

- **Justice**: Each person deserves equal consideration in how to be treated, and the benefits of research should be fairly distributed according to individual need, effort, societal contribution, and merit; Selection of subjects should be fair, and burdens should be allocated equitably across impacted subjects.

- **Respect for law and public interest**: Engage in legal due diligence; Be transparent in methods and results; Be accountable for actions.

The report encourages researchers and engineers to make a stakeholder identification and an identification of their perspectives and considerations. One of these considerations is the presence of malicious actors who can abuse the results of our work and/or harm the individuals whose data we analyze. We must always protect the people whose data we analyze and ensure that we do not cause them any harm. The same applies to the organizations we work with.

Ethics behind data collected from humans

In Europe, one of the main legal frameworks that regulates how we can use the data is the **General Data Protection Regulation (GDPR)** (`https://eur-lex.europa.eu/legal-content/EN/TXT/PDF/?uri=CELEX:32016R0679`). It regulates the scope of handling personal data and puts requirements on the organization to obtain permission to collect, process, and use personal data, as well as requiring organizations to provide individuals with the ability to revoke permissions. The regulation is the most restrictive international regulation that is meant to protect individuals (us) from being abused by companies that have the means and abilities to collect and process data about us.

Although we use a lot of data from GitHub and similar repositories, there are repositories where we also store the data. One of them is Zenodo, which is used increasingly often to store datasets. Its terms of use require us to obtain the right permissions. Here are its terms of use (`https://about.zenodo.org/terms/`):

```
1. Zenodo is an open dissemination research data repository for
the preservation and making available of research, educational and
informational content. Access to Zenodo's content is open to all, for
non-military purposes only.
2. Content may be uploaded free of charge by those without ready
access to an organised data centre.
3. The uploader is exclusively responsible for the content that they
upload to Zenodo and shall indemnify and hold CERN free and harmless
in connection with their use of the service. The uploader shall ensure
that their content is suitable for open dissemination, and that it
complies with these terms and applicable laws, including, but not
limited to, privacy, data protection and intellectual property rights
*. In addition, where data that was originally sensitive personal data
is being uploaded for open dissemination through Zenodo, the uploader
shall ensure that such data is either anonymised to an appropriate
```

```
degree or fully consent cleared **.
4. Access to Zenodo, and all content, is provided on an "as-is"
basis. Users of content ("Users") shall respect applicable license
conditions. Download and use of content from Zenodo does not transfer
any intellectual property rights in the content to the User.
5. Users are exclusively responsible for their use of content, and
shall indemnify and hold CERN free and harmless in connection with
their download and/or use. Hosting and making content available
through Zenodo does not represent any approval or endorsement of such
content by CERN.
6. CERN reserves the right, without notice, at its sole discretion
and without liability, (i) to alter, delete or block access to content
that it deems to be inappropriate or insufficiently protected, and
(ii) to restrict or remove User access where it considers that use of
Zenodo interferes with its operations or violates these Terms of Use
or applicable laws.
7. Unless specified otherwise, Zenodo metadata may be freely reused
under the CC0 waiver.
8. These Terms of Use are subject to change by CERN at any time and
without notice, other than through posting the updated Terms of Use on
the Zenodo website.


* Uploaders considering Zenodo for the storage of unanonymised or
encrypted/unencrypted sensitive personal data are advised to use
bespoke platforms rather than open dissemination services like Zenodo
for sharing their data
** See further the user pages regarding uploading for information on
anonymisation of datasets that contain sensitive personal information.
```

The important part is about the responsibility for content:

- If you upload content, you're solely responsible for it.

- We must ensure that our content is appropriate for public viewing and follows all relevant laws and these terms. This includes laws and terms related to privacy, data protection, and **intellectual property (IP)** rights.

- If we're uploading sensitive personal data, we must make sure it's either appropriately anonymized or we have full consent to share it.

I cannot stress that enough – the information that we provide at Zenodo is our responsibility, so we should make sure that we do not do any harm by providing it open to everyone. If there is non-anonymized data, we should consider other types of storage; for example, storage that requires authentication or access control, hence my next best practice.

> **Best practice #72**
> Any personal data should be stored behind authentication and access control to prevent malicious actors from accessing it.

Although we may have permission to use personal, non-anonymized data, we should keep this kind of data behind authentication. We need to protect the individuals who are behind this data. We also need to use access control and monitoring in case we need to backtrack who accessed the data; for instance, when a mistake was made.

Contracts and legal obligations

To finish the chapter, I would like to take up one last topic. Although there is a lot of data available, we must make sure that we do our due diligence and find out which contracts and obligations apply to us.

Licenses are one type of contract, but not the only one. Almost all universities put contracts and obligations on researchers. These may include the need to ask for permission from ethical review boards or the need to make data available for scrutiny from other researchers.

Professional codes of conduct are another type of obligation; for example, the one from ACM (`https://www.acm.org/code-of-ethics`). These codes of conduct often stem from the Nuremberg Code and require us to ensure that our work is for the good of society.

Finally, when working with commercial organizations, we may need to sign a so-called **non-disclosure agreement (NDA)**. Such agreements are often required to ensure that we do not disclose information without prior permission. They are often mistaken for the need to conceal information, but in most cases, it means that we need to ensure that our reporting of proprietary information does not harm the organization. In most cases, we may need to ensure that our report is about general practices rather than a specific company. If we see deficiencies with our industrial partners, we need to discuss them with our industrial partners and help them to improve – because we need to work for the best of society.

Therefore, I strongly encourage you to find out which codes of conduct, obligations, and contracts are applicable to you.

References

- *Code, N., The Nuremberg Code. Trials of war criminals before the Nuremberg military tribunals under control council law, 1949. 10(1949): p. 181-2.*

- *Wohlin, C. et al., Experimentation in software engineering. 2012: Springer Science & Business Media.*

- *Gold, N.E. and J. Krinke, Ethics in the mining of software repositories. Empirical Software Engineering, 2022. 27(1): p. 17.*

- *Kenneally, E. and D. Dittrich, The Menlo Report: Ethical principles guiding information and communication technology research. Available at SSRN 2445102, 2012.*

15

Ethics in Machine Learning Systems

Ethics involves data acquisition and management and focuses on collecting data, with a particular focus on protecting individuals and organizations from any harm that could be inflicted upon them. However, data is not the only source of bias in **machine learning** (**ML**) systems.

Algorithms and ways of data processing are also prone to introducing bias to the data. Despite our best efforts, some of the steps in data processing may even emphasize the bias and let it spread beyond algorithms and toward other parts of ML-based systems, such as user interfaces or decision-making components.

Therefore, in this chapter, we'll focus on the bias in ML systems. We'll start by exploring sources of bias and briefly discussing these sources. Then, we'll explore ways to spot biases, how to minimize them, and finally how to communicate potential bias to the users of our system.

In this chapter, we're going to cover the following main topics:

- Bias and ML – is it possible to have an objective AI?
- Measuring and monitoring for bias
- Reducing bias
- Developing mechanisms to prevent ML bias from spreading in the entire system

Bias and ML – is it possible to have an objective AI?

In the intertwined domains of ML and software engineering, the allure of data-driven decision-making and predictive modeling is undeniable. These fields, which once operated largely in silos, now converge in numerous applications, from software development tools to automated testing frameworks. However, as we increasingly rely on data and algorithms, a pressing concern emerges: the issue of bias. Bias, in this context, refers to systematic and unfair discrepancies that can manifest in the decisions and predictions of ML models, often stemming from the very data used in software engineering processes.

The sources of bias in software engineering data are multifaceted. They can arise from historical project data, user feedback loops, or even the design and objectives of the software itself. For instance, if a software tool is predominantly tested and refined using feedback from a specific demographic, it might inadvertently underperform or misbehave for users outside that group. Similarly, a defect prediction model might be skewed if trained on data from projects that lack diversity in team composition or coding practices.

The implications of such biases extend beyond mere technical inaccuracies. They can lead to software products that alienate or disadvantage certain user groups, perpetuating and amplifying existing societal inequalities. For example, a development environment might offer suggestions that resonate more with one cultural context than another, or a software recommendation system might favor applications from well-known developers, sidelining newcomers.

Generally, bias is defined to be an inclination or prejudice for, or against, one person or group. In ML, bias is when a model systematically produces prejudiced results. There are several types of bias in ML:

- **Prejudicial bias**: This is a type of bias that is present in the empirical world and made its way into ML models and algorithms – both knowingly and unknowingly. An example is racial bias or gender bias.

- **Measurement bias**: This is a type of bias that is introduced through a systematic error in our measurement instruments. For example, we measure the McCabe complexity of software modules by counting the if/for statements, excluding the while loops.

- **Sampling bias**: This is a type of bias that occurs when our sample does not reflect the real distribution of data. It can be the case that we sample too often or too seldom from a specific class – such a bias in our data will affect inference.

- **Algorithm bias**: This is a type of bias that occurs when we use the wrong algorithm for the task at hand. A wrong algorithm may not generalize well and therefore it may introduce bias into the inference.

- **Confirmation bias**: This is a type of bias that is introduced when we remove/select data points that are aligned with the theoretical notions that we want to capture. By doing this, we introduce the bias that confirms our theory, rather than reflecting the empirical world.

This list is, by no means, exclusive. Bias can be introduced in many ways and through many ways, but it is always our responsibility to identify it, monitor it, and reduce it.

Luckily, there are a few frameworks that can allow us to identify bias – Fair ML, IBM AI Fairness 360, and Microsoft Fairlearn, just to name a few. These frameworks allow us to scrutinize our algorithms and datasets in search of the most common biases.

Donald et al. present a recent overview of methods and tools for reducing bias in software engineering, which includes ML. The important part of that article is that it focuses on use cases, which is important for understanding bias; bias is not something universal but depends on the dataset and the use case of that data. In addition to the sources of bias presented previously, they also recognize that bias is something that can change over time, just as our society changes and just as our data changes. Although Donald et al.'s work is generic, it tends to focus on one of the data types – natural language – and how bias can be present. They provide an overview of tools and techniques that can help identify such phenomena as hateful language.

In this chapter, however, we'll focus on one of the frameworks that is a bit more generic to illustrate how to work with bias in general.

Measuring and monitoring for bias

Let's look at one of these frameworks – IBM AI Fairness 360 (`https://github.com/Trusted-AI/AIF360`). The basis for this framework is the ability to set variables that can be linked to bias and then calculate how different the other variables are. So, let's dive into an example of how to calculate bias for a dataset. Since bias is often associated with gender or similar attributes, we need to use a dataset that contains it. So far in this book, we have not used any dataset that contained this kind of attribute, so we need to find another one.

Let's take the Titanic survival dataset to check if there was any bias in terms of survivability between male and female passengers. First, we need to install the IBM AI Fairness 360 framework:

```
pip install aif360
```

Then, we can start creating a program that will check for bias. We need to import the appropriate libraries and create the data. In this example, we'll create the data of salaries, which is biased toward men:

```
import numpy as np
import pandas as pd
from sklearn.model_selection import train_test_split
from sklearn.preprocessing import StandardScaler
from sklearn.linear_model import LogisticRegression
from aif360.datasets import BinaryLabelDataset
from aif360.metrics import BinaryLabelDatasetMetric
from aif360.algorithms.preprocessing import Reweighing


t i
data = {
    'Age': [25, 45, 35, 50, 23, 30, 40, 28, 38, 48, 27, 37, 47, 26,
36, 46],
    'Income': [50000, 100000, 75000, 120000, 45000, 55000, 95000,
65000, 85000, 110000, 48000, 58000, 98000, 68000, 88000, 105000],
    'Gender': [1, 0, 1, 0, 1, 0, 1, 0, 1, 0, 1, 0, 1, 0, 1, 1],  # 1:
Male, 0: Female
    'Hired': [1, 1, 1, 0, 1, 0, 1, 0, 1, 0, 1, 0, 1, 0, 1, 1]   # 1:
Hired, 0: Not Hired
```

```
}
df = pd.DataFrame(data)
```

This data contains four different attributes – age, income, gender, and whether the person is recommended to be hired or not. It is difficult to spot whether there is a bias between the genders, but let's apply the IBM fairness algorithm to check for that:

```
# Split data into training and testing sets
train, test = train_test_split(df, test_size=0.2, random_state=42)

# Convert dataframes into BinaryLabelDataset format
train_bld = BinaryLabelDataset(df=train, label_names=['Hired'],
protected_attribute_names=['Gender'])
test_bld = BinaryLabelDataset(df=test, label_names=['Hired'],
protected_attribute_names=['Gender'])

# Compute fairness metric on original training dataset
metric_train_bld = BinaryLabelDatasetMetric(train_bld, unprivileged_
groups=[{'Gender': 1}], privileged_groups=[{'Gender': 0}])
print(f'Original training dataset disparity: {metric_train_bld.mean_
difference():.2f}')

# Mitigate bias by reweighing the dataset
RW = Reweighing(unprivileged_groups=[{'Gender': 1}], privileged_
groups=[{'Gender': 0}])
train_bld_transformed = RW.fit_transform(train_bld)

# Compute fairness metric on transformed training dataset
metric_train_bld_transformed = BinaryLabelDatasetMetric(train_
bld_transformed, unprivileged_groups=[{'Gender': 1}], privileged_
groups=[{'Gender': 0}])
print(f'Transformed training dataset disparity: {metric_train_bld_
transformed.mean_difference():.2f}')
```

The preceding code creates a data split and calculates the fairness metric – dataset disparity. The important part of the algorithm is where we set the protected attribute – gender (`protected_attribute_names=['Gender'])`). We manually set the attribute that we think could be prone to bias, which is an important observation. The fairness framework does not set any attributes automatically. Then, we set which values of this attribute indicate the privileged and unprivileged groups – `unprivileged_groups=[{'Gender': 1}]`. Once the code executes, we get an understanding of whether there is bias in the dataset:

```
Original training dataset disparity: 0.86
Transformed training dataset disparity: 0.50
```

This means that the algorithm could reduce the disparity but did not remove it completely. The disparity value of 0.86 means that there is a bias toward the privileged group (in this case males). The value of 0.5 means that the bias is reduced, but it is still far from 0.0, which would indicate the lack of bias. The fact that the bias was reduced and not removed can indicate that there is just too little data to be able to reduce the bias completely.

Therefore, let's take a look at the real dataset, which can contain a bias – the Titanic dataset. The dataset contains protected attributes such as gender and it is significantly larger so that we have a better chance to reduce the bias even more:

```
from aif360.datasets import BinaryLabelDataset
from aif360.metrics import BinaryLabelDatasetMetric
from aif360.algorithms.preprocessing import Reweighing

# Load Titanic dataset
url = "https://web.stanford.edu/class/archive/cs/cs109/cs109.1166/
stuff/titanic.csv"
df = pd.read_csv(url)
```

Now that we have the dataset in place, we can write the script that will calculate the disparity metrics, which quantifies how much difference there is in the data based on the controlled variable:

```
# Preprocess the data
df['Sex'] = df['Sex'].map({'male': 1, 'female': 0})  # Convert 'Sex'
to binary: 1 for male, 0 for female
df.drop(['Name'], axis=1, inplace=True)  # Drop the 'Name' column

# Split data into training and testing sets
train, test = train_test_split(df, test_size=0.2, random_state=42)

# Convert dataframes into BinaryLabelDataset format
train_bld = BinaryLabelDataset(df=train, label_names=['Survived'],
protected_attribute_names=['Sex'])
test_bld = BinaryLabelDataset(df=test, label_names=['Survived'],
protected_attribute_names=['Sex'])

# Compute fairness metric on the original training dataset
metric_train_bld = BinaryLabelDatasetMetric(train_bld, unprivileged_
groups=[{'Sex': 0}], privileged_groups=[{'Sex': 1}])
print(f'Original training dataset disparity: {metric_train_bld.mean_
difference():.2f}')

# Mitigate bias by reweighing the dataset
RW = Reweighing(unprivileged_groups=[{'Sex': 0}], privileged_
groups=[{'Sex': 1}])
train_bld_transformed = RW.fit_transform(train_bld)
```

```
# Compute fairness metric on the transformed training dataset
metric_train_bld_transformed = BinaryLabelDatasetMetric(train_
bld_transformed, unprivileged_groups=[{'Sex': 0}], privileged_
groups=[{'Sex': 1}])
print(f'Transformed training dataset disparity: {metric_train_bld_
transformed.mean_difference():.2f}')
```

First, we need to convert the `'Sex'` column of the DataFrame, `df`, into a binary format: 1 for male and 0 for female. Then, we need to drop the `'Name'` column from the DataFrame as it could be confused with the index. Then, the data is split into training and testing sets using the `train_test_split` function. 20% of the data (`test_size=0.2`) is reserved for testing, and the rest is used for training. `random_state=42` ensures the reproducibility of the split.

Next, we convert the training and testing DataFrames into a `BinaryLabelDataset` format, which is a specific format used by the fairness framework. The target variable (or label) is `'Survived'`, and the protected attribute (that is, the attribute we're concerned about in terms of fairness) is `'Sex'`. The framework considers females (`'Sex': 0`) as the unprivileged group and males (`'Sex': 1`) as the privileged group.

The `mean_difference` method computes the difference in mean outcomes between the privileged and unprivileged groups. A value of 0 indicates perfect fairness, while a non-zero value indicates some disparity. Then, the code uses the `Reweighing` method to mitigate bias in the training dataset. This method assigns weights to the instances in the dataset to ensure fairness. The transformed dataset (`train_bld_transformed`) has these new weights. Then, we calculate the same metric on the transformed dataset. This results in the following output:

```
Original training dataset disparity: 0.57
Transformed training dataset disparity: 0.00
```

This means that the algorithm has balanced the datasets so that the survival rate is the same for male and female passengers. We can now use this dataset to train a model:

```
# Train a classifier (e.g., logistic regression) on the transformed
dataset
scaler = StandardScaler()
X_train = scaler.fit_transform(train_bld_transformed.features)
y_train = train_bld_transformed.labels.ravel()
clf = LogisticRegression().fit(X_train, y_train)

# Test the classifier
X_test = scaler.transform(test_bld.features)
y_test = test_bld.labels.ravel()
y_pred = clf.predict(X_test)

# Evaluate the classifier's performance
```

```
from sklearn.metrics import accuracy_score, classification_report
accuracy = accuracy_score(y_test, y_pred)
print(f"Accuracy: {accuracy:.4f}")
report = classification_report(y_test, y_pred, target_names=["Not
Survived", "Survived"])
print(report)
```

First, we initialize `StandardScaler`. This scaler standardizes features by removing the mean and scaling to unit variance. Then, we transform and standardize the features of the training dataset (`train_bld_transformed.features`) using the `fit_transform` method of the scaler. The standardized features are stored in `X_train`. Then, we extract the labels of the transformed training dataset using the `ravel()` method, resulting in `y_train`. After, we train the logistic regression classifier (`clf`) using the standardized features (`X_train`) and labels (`y_train`).

Then, we standardize the features of the test dataset (`test_bld.features`) using the transform method of the scaler to obtain `X_test`. We do the same with the `y_test` data. We use the trained classifier (`clf`) to make predictions on the standardized test features and store them in `y_pred`.

Finally, we calculate the evaluation scores for the dataset and print the report with accuracy, precision, and recall.

With that, we've come to my best practice related to bias.

> **Best practice #73**
>
> If a dataset contains variables that can be prone to bias, use the disparity metric to get a quick orientation of the data.

It is important to check for bias, although we do not always have access to the variables that we need for its calculations, such as gender or age. If we do not, we should look for attributes that can be correlated with them and check for bias against these attributes.

Other metrics of bias

The dataset disparity metrics that we've used so far are only some of the metrics related to bias. Some of the other metrics that are available in the IBM AI Fairness 360 framework are as follows:

- **True positive rate**: The ratio of true positives conditioned on the protected attribute. This is usually used for classification.

- **False discovery rate**: The difference between the false discovery ratio between the privileged and unprivileged groups in classification tasks.

- **Generalize binary confusion matrix**: The confusion matrix conditioner on the protected attributes in the classification tasks.

- The ratio between privileged and unprivileged instances, which can be used for all kinds of tasks.

There are several metrics in addition to these, but the ones we've covered here illustrate the most important point – or two points. First of all, we can see that there needs to be an attribute, called the protected attribute, which can help us understand the bias. Without such an attribute, the framework cannot do any calculations and therefore it cannot provide any useful feedback for the developers. The second point is the fact that the metrics are based on the unbalance between different groups – privileged and unprivileged – which we define ourselves. We cannot use this framework to discover bias that is hidden.

Hidden biases are biases that are not directly represented by attributes. For example, there are differences in the occupations that men and women have, and therefore the occupation can be such an attribute that is correlated to the gender, but not equal to it. This means that we cannot treat this as a protected attribute, but we need to consider it – basically, there are no occupations that are purely male or purely female occupations, but different occupations have different proportions of men and women.

Developing mechanisms to prevent ML bias from spreading throughout the system

Unfortunately, it is generally not possible to completely remove bias from ML as we often do not have access to the attributes needed to reduce the bias. However, we can reduce the bias and reduce the risk that the bias spreads to the entire system.

Awareness and education are some of the most important measures that we can use to manage bias in software systems. We need to understand the potential sources of bias and their implications. We also need to identify biases related to protected attributes (for example, gender) and identify whether other attributes can be correlated with them (for example, occupation and address). Then, we need to educate our team about the ethical implications of biased models.

Then, we need to diversify our data collection. We must ensure that the data we collect is representative of the population we're to model. To avoid over-representing or under-representing certain groups, we need to ensure that our data collection procedures are scrutinized before they are applied. We also need to monitor for biases in the collected data and reduce them. For example, if we identify a bias in credit scores, we can introduce the data that will prevent this bias from being strengthened by our model.

During data preprocessing, we need to ensure that we handle missing data correctly. Instead of just removing the data points or imputing them with mean values, we should use the right imputation, which takes care of the differences between the privileged and unprivileged groups.

We also need to actively work with the detection of bias. We should use statistical tests to check whether the data distribution is biased toward certain groups, at which point we need to visualize the distributions and identify potential bias. We've already discussed visualization techniques; at this point, we can add that we need to use different symbols for privileged and unprivileged groups to visualize two distributions on the same diagram, for example.

In addition to working with the data, we also need to work with algorithmic fairness, which is when we design the models. We need to set fairness constraints and we need to introduce the attributes that can help us to identify privileged and unprivileged groups. For example, if we know that different occupations have a certain bias toward genders, we need to introduce superficial gender-bias attributes that can help us to create a model that will take that into account and prevent the bias from spreading to other parts of our system. We can also make post-hoc adjustments to the model after training. For example, when predicting a salary, we can adjust that salary based on pre-defined rules after the prediction. That can help to reduce the biases inherent in the model.

We can also use fairness-enhancing interventions, such as the IBM Fairness tools and techniques, which include debiasing, reweighing, and disparate impact removal. This can help us to achieve interpretable models or allow us to use model interpretation tools to understand how decisions are being made. This can help in identifying and rectifying bias.

Finally, we can regularly audit our models for bias and fairness. This includes both automated checks and human reviews. This helps us understand whether there are biases that cannot be captured automatically and that we need to react to.

With that, we have come to my next best practice.

> Best practice #74
> Complement automated bias management with regular audits.

We need to accept the fact that bias is inherent in data, so we need to act accordingly. Instead of relying on algorithms to detect bias, we need to manually monitor for bias and understand it. Therefore, I recommend making regular checks for bias manually. Make classifications and predictions and check whether they strengthen or reduce bias by comparing them to the expected data without bias.

Summary

One of our responsibilities as software engineers is to ensure that we develop software systems that contribute to the greater good of society. We love working with technology development, but the technology needs to be developed responsibly. In this chapter, we looked at the concept of bias in ML and how to work with it. We looked at the IBM Fairness framework, which can assist us in identifying bias. We also learned that automated bias detection is too limited to be able to remove bias from the data completely.

There are more frameworks to explore and more studies and tools are available every day. These frameworks are more specific and provide a means to capture more domain-specific bias – in medicine and advertising. Therefore, my final recommendation in this chapter is to explore the bias frameworks that are specific to the task at hand and for the domain at hand.

References

- *Donald, A., et al., Bias Detection for Customer Interaction Data: A Survey on Datasets, Methods, and Tools. IEEE Access, 2023.*

- *Bellamy, R.K., et al., AI Fairness 360: An extensible toolkit for detecting, understanding, and mitigating unwanted algorithmic bias. arXiv preprint arXiv:1810.01943, 2018.*

- *Zhang, Y., et al. Introduction to AI fairness. In Extended Abstracts of the 2020 CHI Conference on Human Factors in Computing Systems. 2020.*

- *Alves, G., et al. Reducing unintended bias of ml models on tabular and textual data. In 2021 IEEE 8th International Conference on Data Science and Advanced Analytics (DSAA). 2021. IEEE.*

- *Raza, S., D.J. Reji, and C. Ding, Dbias: detecting biases and ensuring fairness in news articles. International Journal of Data Science and Analytics, 2022: p. 1-21.*

16

Integrating ML Systems in Ecosystems

ML systems have gained a lot of popularity for two reasons – their ability to learn from data (which we've explored throughout this book), and their ability to be packaged into web services.

Packaging these ML systems into web services allows us to integrate them into workflows in a very flexible way. Instead of compiling or using dynamically linked libraries, we can deploy ML components that communicate over HTTP protocols using JSON protocols. We have already seen how to use that protocol by using the GPT-3 model that is hosted by OpenAI. In this chapter, we'll explore the possibility of creating a Docker container with a pre-trained ML model, deploying it, and integrating it with other components.

In this chapter, we're going to cover the following main topics:

- ML system of systems – software ecosystems

- Creating web services over ML models using Flask

- Deploying ML models using Docker

- Combining web services into ecosystems

Ecosystems

In the dynamic realm of software engineering, the tools, methodologies, and paradigms are in a constant state of evolution. Among the most influential forces driving this transformation is ML. While ML itself is a marvel of computational prowess, its true genius emerges when integrated into the broader software engineering ecosystems. This chapter delves into the nuances of embedding ML within an ecosystem. Ecosystems are groups of software that work together but are not connected at compile time. A well-known ecosystem is the PyTorch ecosystem, where a set of libraries work together in the context of ML. However, there is much more than that to ML ecosystems in software engineering.

From automated testing systems that learn from each iteration to recommendation engines that adapt to user behaviors, ML is redefining how software is designed, developed, and deployed. However, integrating ML into software engineering is not a mere plug-and-play operation. It demands a rethinking of traditional workflows, a deeper understanding of data-driven decision-making, and a commitment to continuous learning and adaptation.

As we delve deeper into the integration of ML within software engineering, it becomes imperative to discuss two pivotal components that are reshaping the landscape: web services and Docker containers. These technologies, while not exclusive to ML applications, play a crucial role in the seamless deployment and scaling of ML-driven solutions in the software ecosystem.

Web services, especially in the era of microservices architecture, provide a modular approach to building software applications. By encapsulating specific functionalities into distinct services, they allow for greater flexibility and scalability. When combined with ML models, web services can deliver dynamic, real-time responses based on the insights derived from data. For instance, a web service might leverage an ML model to provide personalized content recommendations to users or to detect fraudulent activities in real time.

Docker containers, on the other hand, have revolutionized the way software, including ML models, is packaged and deployed. Containers encapsulate an application along with all its dependencies into a standardized unit, ensuring consistent behavior across different environments. For ML practitioners, this means the painstaking process of setting up environments, managing dependencies, and ensuring compatibility is vastly simplified. Docker containers ensure that an ML model trained on a developer's machine will run with the same efficiency and accuracy on a production server or any other platform.

Furthermore, when web services and Docker containers are combined, they pave the way for ML-driven microservices. Such architectures allow for the rapid deployment of scalable, isolated services that can be updated independently without disrupting the entire system. This is especially valuable in the realm of ML, where models might need frequent updates based on new data or improved algorithms.

In this chapter, we'll learn how to use both technologies to package models and create an ecosystem based on Docker containers. After reading this chapter, we shall have a good understanding of how we can scale up our development by using ML as part of a larger system of systems – ecosystems.

Creating web services over ML models using Flask

In this book, we've mostly focused on training, evaluating, and deploying ML models. However, we did not discuss the need to structure them flexibly. We worked with monolithic software. Monolithic software is characterized by its unified, single code base structure where all the functionalities, from the user interface to data processing, are tightly interwoven and operate as one cohesive unit. This design simplifies initial development and deployment since everything is bundled together and they are compiled together. Any change, however minor, requires the entire application to be rebuilt and redeployed. This makes it problematic when the evolution of contemporary software is fast.

On the other hand, web service-based software, which is often associated with microservices architecture, breaks down the application into smaller, independent services that communicate over the web, typically using protocols such as HTTP and REST. Each service is responsible for a specific functionality and operates independently. This modular approach offers greater flexibility. Services can be scaled, updated, or redeployed individually without affecting the entire system. Moreover, failures in one service don't necessarily bring down the whole application. *Figure 16.1* presents how the difference between these two types of software can be seen:

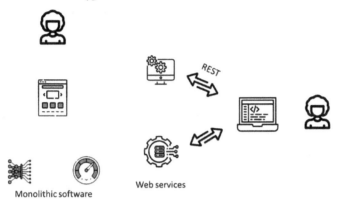

Figure 16.1 – Monolithic software versus web service-based software

On the left-hand side, we have all the components bundled together into one product. Users interact with the product through the user interface. Only one user can interact with the software, and more users require more installations of the software.

On the right-hand side, we have a decentralized architecture where each component is a separate web service. The coordination of these components is done by a thin client. If more users/clients want to use the same services, they can just connect them using the HTTP REST protocol (API).

Here is my first best practice.

> **Best practice #75**
> Use web services (RESTful API) when deploying ML models for production.

Although it takes additional effort to create web services, it is worth using them. They provide a great separation of concerns and asynchronous access and also provide great possibilities for load balancing. We can use different servers to run the same web service and therefore balance the load.

So, let's create the first web service using Flask.

Creating a web service using Flask

Flask is a framework that allows us to provide easy access to internal APIs via the REST interface over HTTP protocol. First, we need to install it:

```
pip install flask
pip install flask-restful
```

Once we've installed the interface, we can write our programs. In this example, our first web service calculates the lines of code and complexity of the program sent to it. The following code fragment exemplifies this:

```python
from fileinput import filename
from flask import *
from radon.complexity import cc_visit
from radon.cli.harvest import CCHarvester


app = Flask(__name__)

# Dictionary to store the metrics for the file submitted
# Metrics: lines of code and McCabe complexity
metrics = {}


def calculate_metrics(file_path):
    with open(file_path, 'r') as file:
        content = file.read()

    # Count lines of code
    lines = len(content.splitlines())

    # Calculate McCabe complexity
    complexity = cc_visit(content)

    # Store the metrics in the dictionary
    metrics[file_path] = {
        'lines_of_code': lines,
        'mccabe_complexity': complexity
    }


@app.route('/')
def main():
    return render_template("index.html")
```

```
@app.route('/success', methods=['POST'])
def success():
    if request.method == 'POST':
        f = request.files['file']

        # Save the file to the server
        file_path = f.filename
        f.save(file_path)

        # Calculate metrics for the file
        calculate_metrics(file_path)

        # Return the metrics for the file
        return metrics[file_path]


@app.route('/metrics', methods=['GET'])
def get_metrics():
    if request.method == 'GET':
        return metrics

if __name__ == '__main__':
    app.run(host='0.0.0.0', debug=True)
```

First, the code requires a few imports and then initializes the application in `app = Flask(__name__)`. Then, it creates the routes – that is, the places where the program will be able to communicate via the REST API:

- `@app.route('/')`: This is a decorator that defines a route for the root URL (`"/"`). When users access the root URL, it renders the `"index.html"` template.

- `@app.route('/success', methods=['POST'])`: This decorator defines a route for the `"/success"` URL, which expects HTTP POST requests. This route is used to handle file uploads, count lines of code, and calculate McCabe complexity.

- `@app.route('/metrics', methods=['GET'])`: This decorator defines a route for the `"/metrics"` URL, which expects HTTP GET requests. It is used to retrieve and display the metrics.

- `def main()`: This function is associated with the root (`"/"`) route. It returns an HTML template called `"index.html"` when users access the root URL.

- `def success()`: This function is associated with the `"/success"` route, which handles file uploads:

  - It checks if the request method is POST

  - It saves the uploaded file to the server

  - It counts the lines of code in the uploaded file

  - It calculates the McCabe complexity using the radon library

  - It stores the metrics (lines of code and McCabe complexity) in the metrics dictionary

  - It returns the metrics for the uploaded file as a JSON response

- `def get_metrics()`: This function is associated with the `/metrics` route:

  - It checks if the request method is GET.

  - It returns the entire metrics dictionary as a JSON response. This is used for debugging purposes to see the metrics for all files that are uploaded during the session.

- `if __name__ == '__main__':`: This block ensures that the web application is only run when this script is executed directly (not when it's imported as a module).

- `app.run(host='0.0.0.0', debug=True)`: This starts the Flask application in debug mode, allowing you to see detailed error messages during development.

Then, the application is executed – `app.run(debug=True)` starts the Flask application with debugging mode enabled. This means that any changes made to the code will automatically reload the server, and any errors will be displayed in the browser. Once we execute it, the following web page appears (note that the code of the page has to be in the `templates` subfolder and should contain the following code):

```
<html>
<head>
    <title>Machine learning best practices for software engineers:
Chapter 16 - Upload a file to make predictions</title>
</head>
<body>
    <h1>Machine learning best practices for software engineers -
Chapter 16</h1>

    <p>This page allows to upload a file to a web service that has
been written using Flask. The web application behind this interface
calculates metrics that are important for the predictions. It returns
a JSON string with the metrics.  </p>

    <p>We need another web app that contains the model in order to
```

```
actually obtain predictions if the file can contain defects. </p>

    <h1>Upload a file to make predictions</h1>
    <p>The file should be a .c or .cpp file</p>

    <form action = "/success" method = "post" enctype="multipart/form-
data">
        <input type="file" name="file" />
        <input type = "submit" value="Upload">
    </form>

    <p>Disclaimer: the container saves the file it its local folder,
so don't send any sensitive files for analysis.</p>

    <p>This is a research prototype</p>
</body>
</html>
```

The page contains a simple form that allows us to upload the file to the server:

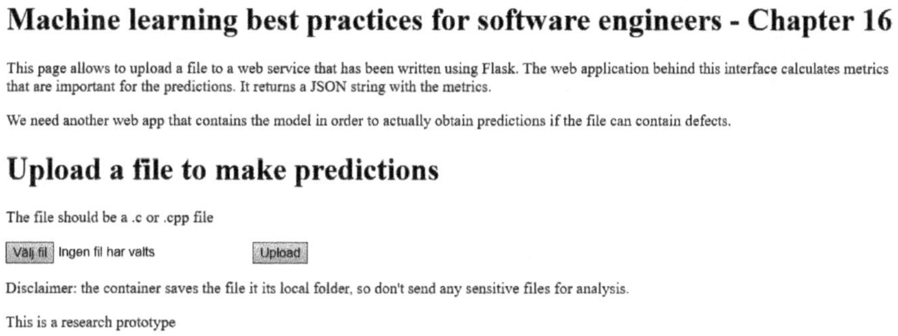

Machine learning best practices for software engineers - Chapter 16

This page allows to upload a file to a web service that has been written using Flask. The web application behind this interface calculates metrics that are important for the predictions. It returns a JSON string with the metrics.

We need another web app that contains the model in order to actually obtain predictions if the file can contain defects.

Upload a file to make predictions

The file should be a .c or .cpp file

[Välj fil] Ingen fil har valts   [Upload]

Disclaimer: the container saves the file it its local folder, so don't send any sensitive files for analysis.

This is a research prototype

Figure 16.2 – The web page where we can send the file to calculate lines of code.
This is only one way of sending this information to the web service

After uploading the file, we get the results:

```
{
    "main.py": 45
}
```

Figure 16.3 – Results of calculating the lines of code

The majority of the transmission is done by the Flask framework, which makes the development a really pleasant experience. However, just counting lines of code and complexity is not a great ML model. Therefore, we need to create another web service with the code of the ML model itself.

Therefore, my next best practice is about dual interfaces.

Best practice #76

Use both a website and an API for the web services.

Although we can always design web services so that they only accept JSON/REST calls, we should try to provide different interfaces. The web interface (presented previously) allows us to test the web service and even send data to it without the need to write a separate program.

Creating a web service that contains a pre-trained ML model

The code of the ML model web service follows the same template. It uses the Flask framework to provide the REST API of the web service for the software. Here is the code fragment that shows this web service:

```
#
# This is a flask web service to make predictions on the data
# that is sent to it. It is meant to be used with the measurement
instrument
#
from flask import *
from joblib import load
import pandas as pd

app = Flask(__name__)    # create an app instance

# entry point where we send JSON with two parameters:
# LOC and MCC
# and make prediction using make_prediction method
@app.route('/predict/<loc>/<mcc>')
def predict(loc,mcc):
    return {'Defect': make_prediction(loc, mcc)}

@app.route('/')
def hello():
    return 'Welcome to the predictor! You need to send a GET request
with two parameters: LOC (lines of code) and MCC (McCabe complexity))'

# the main method for making the prediction
```

```python
# using the model that is stored in the joblib file
def make_prediction(loc, mcc):
    # now read the model from the joblib file
    # and predict the defects for the X_test data

    dt = load('dt.joblib')

    # input data to the model
    input = {'LOC': loc,
             'MCC': mcc}

    # convert input data into dataframe
    X_pred = pd.DataFrame(input, index=[0])

    # make prediction
    y_pred = dt.predict(X_pred)

    # return the prediction
    # as an integer
    return int(y_pred[0])

# run the application
if __name__ == '__main__':
    app.run(debug=True)
```

The main entry point to this web service takes two parameters: `@app.route('/predict/<loc>/<mcc>')`. It uses these two parameters as parameters of the method that instantiates the models and uses it to make a prediction – `make_prediction(loc, mcc)`. The `make_prediction` method reads a model from a `joblib` file and uses it to predict whether the module will contain a defect or not. I use `joblib` for this model as it is based on a NumPy array. However, if a model is based on a Python object (for example, when it is an estimator from a scikit-learn library), then it is better to use pickle instead of `joblib`. It returns the JSON string containing the result. *Figure 16.4* illustrates how we can use a web browser to invoke this web service – not the address bar:

```
{
    "Defect": 1
}
```

Figure 16.4 – Using the prediction endpoint to get the predicted number of defects in this module

The address bar sends the parameters to the model and the response is a JSON string that says that this module will, most probably, contain a defect. Well, this is not a surprise as we say that the module has 10 lines of code and a complexity of 100 (unrealistic, but possible).

These two web services already give us an example of how powerful the REST API can be. Now, let's learn how to package that with Docker so that we can deploy these web services even more easily.

Deploying ML models using Docker

To create a Docker container with our newly created web service (or two of them), we need to install Docker on our system. Once we've installed Docker, we can use it to compile the container.

The crucial part of packaging the web service into the Docker container is the Dockerfile. It is a recipe for how to assemble the container and how to start it. If you're interested, I've suggested a good book about Docker containers in the *Further reading* section so that you can learn more about how to create more advanced components than the ones in this book.

In our example, we need two containers. The first one will be the container for the measurement instrument. The code for that container is as follows:

```
FROM alpine:latest
RUN apk update
RUN apk add py-pip
RUN apk add --no-cache python3-dev
RUN pip install --upgrade pip
WORKDIR /app
COPY . /app
RUN pip --no-cache-dir install -r requirements.txt
CMD ["python3", "main.py"]
```

This Dockerfile is setting up an Alpine Linux-based environment, installing Python and the necessary development packages, copying your application code into the image, and then running the Python script as the default command when the container starts. It's a common pattern for creating Docker images for Python applications. Let's take a closer look:

1. `FROM alpine:latest`: This line specifies the base image for the Docker image. In this case, it uses the Alpine Linux distribution, which is a lightweight and minimalistic distribution that's often used in Docker containers. `latest` refers to the latest version of the Alpine image available on Docker Hub.

2. `RUN apk update`: This command updates the package index of the Alpine Linux package manager (`apk`) to ensure it has the latest information about available packages.

3. `RUN apk add py-pip`: Here, it installs the `py-pip` package, which is the package manager for Python packages. This step is necessary to be able to install Python packages using `pip`.

4. `RUN apk add --no-cache python3-dev`: This installs the `python3-dev` package, which provides development files for Python. These development files are often needed when compiling or building Python packages that have native code extensions.

5. `RUN pip install --upgrade pip`: This upgrades the `pip` package manager to the latest version.

6. `WORKDIR /app`: This sets the working directory for the subsequent commands to `/app`. This directory is where the application code will be copied, and it becomes the default directory for running commands.

7. `COPY . /app`: This copies the contents of the current directory (where the Dockerfile is located) into the `/app` directory in the Docker image. This typically includes the application code, including `requirements.txt`.

8. `RUN pip --no-cache-dir install -r requirements.txt`: This installs Python dependencies specified in the `requirements.txt` file. The `--no-cache-dir` flag is used to ensure that no cache is used during the installation, which can help reduce the size of the Docker image.

9. `CMD ["python3", "main.py"]`: This specifies the default command to run when a container is started from this image. In this case, it runs the `main.py` Python script using `python3`. This is the command that will be executed when we run a container based on this Docker image.

In *step 8*, we need the `requirements.txt` file. In this case, the file does not have to be too complex – it needs to use the same imports as the web service script:

```
flask
flask-restful
```

Now, we are ready to compile the Docker container. We can do that with the following command from the command line:

```
docker build -t measurementinstrument .
```

Once the compilation process is complete, we can start the container:

```
docker run -t -p 5000:5000 measurementinstrument
```

The preceding command tells the Docker environment that we want to start the container, `measurementinstrument`, and map the port of the web service (`5000`) to the same port in `localmachine`. Now, if we navigate to the address, we can upload the file just like we could when the web service was running without the Docker containers.

> **Best practice #77**
> Dockerize your web services for both version control and portability.

Using Docker is one way we can ensure the portability of our web services. Once we package our web service into the container, we can be sure that it will behave the same on every system that is capable of running Docker. This makes our lives much easier even more than using `requirements.txt` files to set up Python environments.

Once we have the container with the measurement instrument, we can package the second web service – with the prediction model – into another web service. The following Dockerfile does this:

```
FROM ubuntu:latest
RUN apt update && apt install python3 python3-pip -y
WORKDIR /app
COPY . /app
RUN pip --no-cache-dir install -q -r requirements.txt
CMD ["python3", "main.py"]
```

This Dockerfile sets up an Ubuntu-based environment, installs Python 3 and `pip`, copies your application code into the image, installs Python dependencies from `requirements.txt`, and then runs the Python script as the default command when the container starts. Please note that we use Ubuntu here and not Alpine Linux. This is no accident. There is no scikit-learn package for Alpine Linux, so we need to use Ubuntu (for which that Python package is available):

- `FROM ubuntu:latest`: This line specifies the base image for the Docker image. In this case, it uses the latest version of the Ubuntu Linux distribution as the base image. `latest` refers to the latest version of the Ubuntu image available on Docker Hub.

- `RUN apt update && apt install python3 python3-pip -y`: This command is used to update the package index of the Ubuntu package manager (`apt`) and then install Python 3 and Python 3 `pip`. The `-y` flag is used to automatically answer "yes" to any prompts during the installation process.

- `WORKDIR /app`: This sets the working directory for the subsequent commands to `/app`. This directory is where the application code will be copied, and it becomes the default directory for running commands.

- `COPY . /app`: This copies the contents of the current directory (where the Dockerfile is located) into the `/app` directory in the Docker image.

- `RUN pip --no-cache-dir install -q -r requirements.txt`: This installs Python dependencies specified in the `requirements.txt` file using pip. The flags that are used here are as follows:

  - `--no-cache-dir`: This ensures that no cache is used during the installation, which can help reduce the size of the Docker image

  - `-q`: This flag runs `pip` in quiet mode, meaning it will produce less output, which can make the Docker build process less verbose

- CMD ["python3", "main.py"]: This specifies the default command to run when a container is started from this image. In this case, it runs the main.py Python script using python3. This is the command that will be executed when we run a container based on this Docker image.

In this case, the requirements code is a bit longer, although not extremely complex:

```
scikit-learn
scipy
flask
flask-restful
joblib
pandas
numpy
```

We compile the Docker container with a similar command:

```
docker build -t predictor .
```

We execute it with a similar command:

```
docker run -t -p 5001:5000 predictor
```

Now, we should be able to use the same browser commands to connect. Please note that we use a different port so that this new web service does not collide with the previous one.

Combining web services into ecosystems

Now, let's develop the software that will connect these two web services. For this, we'll create a new file that will send one file to the first web service, get the data, and then send it to the second web service to make predictions:

```
import requests

# URL of the Flask web service for file upload
upload_url = 'http://localhost:5000/success'  # Replace with the
actual URL

# URL of the Flask web service for predictions
prediction_url = 'http://localhost:5001/predict/'  # Replace with the
actual URL


def upload_file_and_get_metrics(file_path):
    try:
```

```
        # Open and read the file
        with open(file_path, 'rb') as file:
            # Create a dictionary to hold the file data
            files = {'file': (file.name, file)}

            # Send a POST request with the file to the upload URL
            response = requests.post(upload_url, files=files)
            response.raise_for_status()

            # Parse the JSON response
            json_result = response.json()

            # Extract LOC and mccabe_complexity from the JSON result
            loc = json_result.get('lines_of_code')
            mccabe_complexity = json_result.get('mccabe_complexity')
[0][-1]

            if loc is not None and mccabe_complexity is not None:
                print(f'LOC: [3], McCabe Complexity: {mccabe_
complexity}')
                return loc, mccabe_complexity
            else:
                print('LOC or McCabe Complexity not found in JSON
result.')

    except Exception as e:
        print(f'Error: {e}')

def send_metrics_for_prediction(loc, mcc):
    try:
        # Create the URL for making predictions
        predict_url = f'{prediction_url}[3]/[4]'

        # Send a GET request to the prediction web service
        response = requests.get(predict_url)
        response.raise_for_status()

        # Parse the JSON response to get the prediction
        prediction = response.json().get('Defect')

        print(f'Prediction: {prediction}')

    except Exception as e:
        print(f'Error: {e}')
```

```
if __name__ == '__main__':
    # Specify the file path you want to upload
    file_path = './main.py'  # Replace with the actual file path

    # Upload the specified file and get LOC and McCabe Complexity
    loc, mcc = upload_file_and_get_metrics(file_path)

    send_metrics_for_prediction(loc, mcc)
```

This code demonstrates how to upload a file to a Flask web service to obtain metrics and then send those metrics to another Flask web service for making predictions. It uses the `requests` library to handle HTTP requests and JSON responses between the two services:

- `import requests`: This line imports the `requests` library, which is used to send HTTP requests to web services.

- `upload_url`: This variable stores the URL of the Flask web service for file upload.

- `prediction_url`: This variable stores the URL of the Flask web service for predictions.

- `upload_file_and_get_metrics`:

  - This function takes `file_path` as input, which should be the path to the file you want to upload and obtain metrics for

  - It sends a POST request to `upload_url` to upload the specified file

  - After uploading the file, it parses the JSON response received from the file upload service

  - It extracts the `"lines_of_code"` and `"mccabe_complexity"` fields from the JSON response

  - The extracted metrics are printed, and the function returns them

- `send_metrics_for_prediction`:

  - This function takes `loc` (lines of code) and `mcc` (McCabe complexity) values as input

  - It constructs the URL for making predictions by appending the `loc` and `mcc` values to `prediction_url`

  - It sends a GET request to the prediction service using the constructed URL

  - After receiving the prediction, it parses the JSON response to obtain the `"Defect"` value

  - The prediction is printed to the console

- `if __name__ == '__main__':` This block specifies the file path (`file_path`) that you want to upload and obtain metrics for. It calls the `upload_file_and_get_metrics` function to upload the file and obtain the metrics (lines of code and McCabe complexity). Then, it calls the `send_metrics_for_prediction` function to send these metrics for prediction and prints the prediction.

This program shows that we can package our model into a web service (with or without a container) and then use it, just like *Figure 16.1* suggested. This way of designing the entire system allows us to make the software more scalable and robust. Depending on the usage scenario, we can adapt the web services and deploy them on several different servers for scalability and load balancing.

Summary

In this chapter, we learned how to deploy ML models using web services and Docker. Although we only deployed two web services, we can see that it can become an ecosystem for ML. By separating predictions and measurements, we can separate the computational-heavy workloads (prediction) and the data collection parts of the pipeline. Since the model can be deployed on any server, we can reuse the servers and therefore reduce the energy consumption of these models.

With that, we have come to was last technical chapter of this book. In the next chapter, we'll take a look at the newest trends in ML and peer into our crystal ball to predict, or at least guess, the future.

References

- *Masse, M., REST API design rulebook: designing consistent RESTful web service interfaces. 2011: "O'Reilly Media, Inc.".*

- *Raj, P., J.S. Chelladhurai, and V. Singh, Learning Docker. 2015: Packt Publishing Ltd.*

- *Staron, M., et al. Robust Machine Learning in Critical Care—Software Engineering and Medical Perspectives. In 2021 IEEE/ACM 1st Workshop on AI Engineering-Software Engineering for AI (WAIN). 2021. IEEE.*

- *McCabe, T.J., A complexity measure. IEEE Transactions on Software Engineering, 1976(4): p. 308-320.*

17

Summary and Where to Go Next

This is the last chapter of this book. We've learned a lot – starting with understanding the differences between traditional and machine learning-based software. We've learned how to handle data and how to work with algorithms. We've also looked at how to deploy models and how to work ethically with machine learning. In this chapter, we'll summarize the best practices and try to get a glimpse of future developments in the area of machine learning overlapped with software engineering.

In this chapter, we're going to cover the following main topics:

- To know where we are going, we need to know where we've been

- Best practices

- Current developments

- My view on the future

To know where we're going, we need to know where we've been

My journey with computers started in the early 1990s, with Atari 800XL. Once I got my hands on that computer, I was amazed by the sheer fact that it could do what I told it to do. My first program was, of course, a program in BASIC:

```
10 PRINT Hello world!
20 GOTO 10
```

It's neither well-formed nor a very useful program, but that was all I could do at the time. This first program shaped my entire career as it sparked my interest in software development. Later on, during my professional career, I realized that professional software engineering is much more than just writing source code and compiling it. The programs need to be well-formed, well-documented, well-designed, and well-tested (among many other things). This observation, in turn, shaped my view on software engineering as a discipline that can turn homebrewed software into a piece of art that can be used over long periods if well maintained. That was in the early 2000s. Around 2015, I saw a potential in machine learning that also sparked my interest. The first project that I took part in was done by a colleague of mine who showed me how a random forest classifier can interpret programming language code and make inferences about defects. Fast forward to today, we have transformers, diffusers, autoencoders, and hybrid networks that can do amazing things with data.

However, we need to move from AI development to AI engineering, as we did from software development to software engineering. That brings us to this book, where I showed over 70 best practices in AI engineering. This book has taken us from the basic concepts of machine learning and AI in *Part I*, through handling data in *Part II* to algorithms and models in *Part III*. We also explored the ethical and legal aspects of designing, using, and deploying machine learning systems in *Part IV*.

Best practices

The first part of this book contains significantly more best practices, which is because these best practices relate to engineering software, designing it, and making crucial decisions about machine learning – for example, the first best practice tells us when to use (and not to use) machine learning.

As this part of this book was about the *machine learning landscape in software engineering*, we'll discuss different types of models and data and show how they come together.

The list of best practices from the first part of this book is presented in *Table 17.1*:

ID	Best Practice
1	Use machine learning algorithms when your problem is focused on data, not on the algorithm.
2	Before you start developing a machine learning system, do due diligence and identify the right group of algorithms to use.

ID	Best Practice
3	If your software is safety-critical, make sure that you can design mechanisms to prevent hazards caused by the probabilistic nature of machine learning.
4	Test the machine learning software as an addition to the typical train-validation-evaluation process of machine learning model development.
5	When designing machine learning software, focus on your data and the problem to solve first and on the algorithm second.
6	Use data imputation only when you know which properties of data you need to strengthen.
7	Once you have explored the problem to solve and understood the data available, decide whether you want to use supervised, self-supervised, unsupervised, or reinforcement learning algorithms.
8	Choose the data validation attributes that are the most relevant for your system.
9	Use GridSearch and other algorithms after you have explored the parameter search space manually.
10	Always include monitoring mechanisms in your machine learning systems.
11	Choose the right database for your data – look at this from the perspective of the data, not the system.
12	Use cloud infrastructure if you can as it saves resources and reduces the need for specialized competence.
13	Decide on your production environment early and align your process with that environment.
14	Design the entire software system based on the task that you need to solve, not only the machine learning model.
15	Downsize the size of your images and use as few colors as possible to reduce the computational complexity of your system.
16	Use a reference dataset for benchmarking whenever you can.
17	Whenever possible, use models that are already pre-trained for specific tasks.
18	Visualize your raw data to get an understanding of patterns in your data.
19	Visualize your data when it has been turned into features to monitor if the same patterns are still observable.
20	Only use the necessary information as input to machine learning models. Too much information may require additional processing and pose unnecessary requirements on the system.
21	Use bounding boxes in the data when the task requires detecting and tracking objects.
22	Use semantic maps when you need to get the context of the image or you need details of a specific area.
23	Use a pre-trained embedding model such as GPT-3 or an existing BERT model to vectorize your text.
24	Use role labels when designing software that needs to provide grounded decisions.

ID	Best Practice
25	Identify the origin of the data used in your software and create your data processing pipeline accordingly.
26	Extract as much data as you need and store it locally to reduce the disturbances for the software engineers who will use the tool for their work.
27	When accessing data from public repositories, please check the licenses and ensure you acknowledge the contribution of the community that created the analyzed code.
28	The best strategy to reduce the impact of noise on machine learning classifiers is to remove the noisy data points.
29	Balance the number of features with the number of data points. Having more features is not always better.
30	Use `KNNImputer` for data for classification tasks and `IterativeImputer` for data for regression tasks.
31	Use the random forest classifier to reduce the attribute noise as it provides very good performance.
32	Retain the original distribution of the data as much as possible as it reflects the empirical observations.
33	In large-scale software systems, if possible, rely on the machine learning models to handle noise in the data.

Table 17.1: Best practices in the first part of this book

The second part of this book was about *data acquisition and management*. We focused on machine learning from the perspective of the data – how to get orientation in the data, what kind of data exists, and how to process it.

Therefore, the following best practices are related to data:

ID	Best Practice
34	When working with numerical data, visualize it first, starting with the summary views of the data.
35	When visualizing data on the aggregate level, focus on the strength of relationships and connections between the values.
36	Diving deeper into individual analyses should be guided by the machine learning task at hand.
37	When visualizing the metadata for images, make sure you visualize the images themselves.
38	Summary statistics for text data help with performing a sanity check of the data.
39	Use feature engineering techniques if the data is complex but the task is simple – for example, creating a classification model.
40	Use PCA if the data is somehow linear and on similar scales.

ID	Best Practice
41	Use t-SNE if you do not know the properties of the data and the dataset is large (>1,000 data points).
42	Use autoencoders for numerical data when the dataset is really large since autoencoders are complex and require a lot of data for training.
43	Start with a small number of neurons in the bottleneck – usually one third of the number of columns. If the autoencoder does not learn, increase the number gradually.
44	Use tokenizers for large language models such as BERT and word embeddings for simple tasks.
45	Use BoW tokenizers together with dictionaries when your task requires a pre-defined set of words.
46	Use the WordPiece tokenizer as your first choice.
47	Use BPE when you're working with large language models and large corpora of text.
48	Use the sentence piece tokenizer when whitespaces are important.
49	Use word embeddings (FastText) as a go-to model for designing classifiers or text.

Table 17.2: Best practices in the second part of this book

Once we worked with the data, we focused on algorithms, which comprise the other part of machine learning.

In the third part of this book, *Designing and Developing Machine Learning Systems*, our focus was on the software. We started by exploring which algorithms exist and how to choose the best ones. Starting from AutoML and working with Hugging Face provided us with a good platform to learn how to train and deploy machine learning systems.

Here are its best practices:

ID	Best Practice
50	Use AutoML as the initial choice when training the classical machine learning models.
51	Use pre-trained models from Hugging Face or TensorFlow Hub to start with.
52	Work with the pre-trained networks to identify their limitations and then train the network on your dataset.
53	Instead of looking for more complex models, create a smarter pipeline.
54	If you want to understand your numerical data, use models that provide explainability, such as decision tree or random forest.
55	The best models are those that capture the empirical phenomena in the data.
56	Simple, but explainable, models can often capture the data in a good way.
57	Always make sure that the data points in both the train and test sets are separate.
58	Use NVidia CUDA (accelerated computing) for training advanced models such as BERT, GPT-3, and Autoencoders.

ID	Best Practice
59	In addition to monitoring the loss, make sure you visualize the actual results of the generation.
60	Check the output of generative AI models so that it does not break the entire system or does not provide unethical responses.
61	The model card should contain information about how the model was trained, how to use it, which tasks it supports, and how to reference the model.
62	Experiment with different models to find the best pipeline.
63	Use a professional testing framework such as Pytest.
64	Set up your test infrastructure based on your training data.
65	Treat models as units and prepare unit tests for them accordingly.
66	Identify the key aspects of the machine learning deployment and monitor these aspects accordingly.
67	Focus on the user task when designing the user interface of the machine learning model.
68	Prepare your models for web deployment.
69	Try to work with in-memory databases and dump them to disk very often.

Table 17.3: Best practices in the third part of this book

In the last part of this book, *Ethical Aspects of Data Management and Machine Learning System Development*, we explored the challenges that relate to machine learning from the perspective of ethics and legal aspects. We offered certain solutions, but they do not replace human judgment and human intelligence. In the final chapter, we returned to a more technical part and taught you how to work with ecosystems, closing this book by opening up new alleys – microservices and Docker:

ID	Best Practice
70	If your model/software aims to help with the daily tasks of your users, make sure that you develop it as an add-in.
71	If you create your own data for non-commercial purposes, use one of the permissive derivative licenses that limit your liability.
72	Limit yourself to studying source code and other artifacts, and only use personal data with the consent of the subjects.
73	Any personal data should be stored behind authentication and access control to prevent malicious actors from accessing it.
74	If a dataset contains variables that can be prone to bias, use the disparity metric to get a quick orientation about the data.
75	Complement automated bias management with regular audits.
76	Use web services when deploying machine learning models to production.
77	Use both a website and a Python API for the web services.
78	Dockerize your web services for both version control and portability.

Table 17.4: Best practices in the fourth part of this book

The best practices presented in this book are based on my experiences in designing, developing, testing, and deploying machine learning systems. These best practices are not exhaustive, and we could find more, depending on our interests.

To look for these best practices, I strongly suggest taking a look at the research publications that appear continuously in this area. In particular, I recommend following the main conferences in AI – that is, NeurIPS and AAAI – as well as the main conferences in software engineering – that is, ICSE, ASE, and ESEC/FSE.

Current developments

At the time of writing this book, the Technology Innovation Institute (`https://www.tii.ae/`) has just released its largest model – Falcon 170B. It is the largest fully open source model that is similar to the GPT-3.5 model. It shows the current direction of the research in large language models.

Although GPT-4 exists, which is larger by a factor of 1,000, we can develop very good software with moderately large models such as GPT-3.5. This brings us to some of the current topics that we, as a community, need to discuss. One of them is the energy sustainability of these models. Falcon-170B requires 400 GB of RAM (eight times that of an Nvidia A100 GPU) to execute (according to Hugging Face). We do not know how much hardware the GPT-4 model needs. The amount of electricity that it takes and the resources that it uses must be on par with what we get as value from that model.

We also approach limits to the conventional computational power when it comes to machine learning, which brings us to quantum computing (Cerezo, Verdon et al. 2022). However, although the idea of embedding machine learning in quantum computing formalism is quite appealing, it is not without its challenges. Machine learning systems/models need to be re-designed to fit the new paradigm and the new paradigm needs to be more accessible for AI engineers and software engineers.

When it comes to algorithms, it is *swarming* that gets increasingly more attention, mostly due to the attention behind the **Internet of Things (IoT)** (Rosenberg 2016, Rajeesh Kumar, Jaya Lakshmi, et al. 2023). With the support of technologies such as 5G and 6G in telecommunication, these developments are clearly on the rise and will continue to rise. However, one of the current challenges is fragmenting the protocols and technologies, which leads to challenges related to cybersecurity and attacks on critical infrastructure.

Yet another alley of research that is currently growing is the area of *graph neural networks* (Gao, Zheng, et al. 2023). As the current architectures are limited to linear data or data or data that can be linearized, they cannot handle certain tasks. Programming language models are one example of such an area – although they are capable of many tasks, they cannot consider execution graphs or similar elements at the moment. Therefore, the graph neural models are seen to be the next development, although the main challenge to address there is related to the computational power needed for training and inference in the graph neural models.

At the same time, increasingly more studies are tackling the development of models and softer aspects of it, such as ethics, explainability, and their impact on society at large (Meskó and Topol 2023). Medicine, large language models, and the digitalization of society are just a few areas where intensive standardization of the work is ongoing.

My view on the future

Based on my observations of the machine learning landscape today, with a particular focus on software engineering, I can see a few trends.

Language models will get better at completing software engineering tasks, such as requirements, testing, and documenting. This means that software engineers will be able to focus on their core work – engineering software – rather than on tedious, repetitive tasks. We will see models that will test software, document it, explain it, and maybe even repair it. The latest advancements in this field are very promising.

Hybrid models will be more popular. Combining symbolic analysis and neural networks will gain traction and be able to assist us in finding advanced vulnerabilities in software, as well as identifying them before they are exploited. This will make our software more robust and more resilient over time.

Large models and the availability of significant computational power will help us also to detect anomalies in software operations. By analyzing logs, we will be able to predict aging and we will be able to direct maintenance effort. Again, we will be able to focus on software engineering rather than on analysis.

And last, but not least, hybrid models will help us with design tasks. Combining image generation and other types of methods will help us in designing software – its architecture and detailed design, and even predicting the operational performance of the software.

Final remarks

I hope that this book was interesting for you and that you have gained new knowledge in machine learning, AI, and engineering software. I hope that you can use this book as a reference and that you will use the associated code to create new products. I would also greatly appreciate it if you could let me know if you liked it, connect via LinkedIn, and contribute to this book's GitHub repository. I will monitor it and integrate all pull requests that you may have.

Before we part, I have one last best practice.

> **Best practice #79**
> Never stop learning.

Take this from a university professor. The field of machine learning grows quickly, with new models being introduced almost every week. Make sure that you observe the scientific publications in this area and commercial developments. This will help you keep your knowledge up-to-date and help you advance in your professional career.

References

- *Cerezo, M., G. Verdon, H.-Y. Huang, L. Cincio and P. J. Coles (2022). Challenges and opportunities in quantum machine learning. Nature Computational Science 2(9): 567-576.*

- *Gao, C., Y. Zheng, N. Li, Y. Li, Y. Qin, J. Piao, Y. Quan, J. Chang, D. Jin and X. He (2023). A survey of graph neural networks for recommender systems: Challenges, methods, and directions. ACM Transactions on Recommender Systems 1(1): 1-51.*

- *Meskó, B. and E. J. Topol (2023). The imperative for regulatory oversight of large language models (or generative AI) in healthcare. npj Digital Medicine 6(1): 120.*

- *Rajeesh Kumar, N. V., N. Jaya Lakshmi, B. Mallala and V. Jadhav (2023). Secure trust aware multi-objective routing protocol based on battle competitive swarm optimization in IoT. Artificial Intelligence Review.*

- *Rosenberg, L. (2016). Artificial Swarm Intelligence, a Human-in-the-loop approach to AI. Proceedings of the AAAI conference on artificial intelligence.*

Index

www.packtpub.com

Subscribe to our online digital library for full access to over 7,000 books and videos, as well as industry leading tools to help you plan your personal development and advance your career. For more information, please visit our website.

Why subscribe?

- Spend less time learning and more time coding with practical eBooks and Videos from over 4,000 industry professionals

- Improve your learning with Skill Plans built especially for you

- Get a free eBook or video every month

- Fully searchable for easy access to vital information

- Copy and paste, print, and bookmark content

Did you know that Packt offers eBook versions of every book published, with PDF and ePub files available? You can upgrade to the eBook version at www.packtpub.com and as a print book customer, you are entitled to a discount on the eBook copy. Get in touch with us at customercare@packtpub.com for more details.

At www.packtpub.com, you can also read a collection of free technical articles, sign up for a range of free newsletters, and receive exclusive discounts and offers on Packt books and eBooks.

Other Books You May Enjoy

If you enjoyed this book, you may be interested in these other books by Packt:

Practical Machine Learning on Databricks

Debu Sinha

ISBN: 978-1-80181-203-0

- Transition smoothly from DIY setups to databricks
- Master AutoML for quick ML experiment setup
- Automate model retraining and deployment
- Leverage databricks feature store for data prep
- Use MLflow for effective experiment tracking
- Gain practical insights for scalable ML solutions
- Find out how to handle model drifts in production environments

Machine Learning for Imbalanced Data

Kumar Abhishek, Dr. Mounir Abdelaziz

ISBN: 978-1-80107-083-6

- Use imbalanced data in your machine learning models effectively

- Explore the metrics used when classes are imbalanced

- Understand how and when to apply various sampling methods such as over-sampling and under-sampling

- Apply data-based, algorithm-based, and hybrid approaches to deal with class imbalance

- Combine and choose from various options for data balancing while avoiding common pitfalls

- Understand the concepts of model calibration and threshold adjustment in the context of dealing with imbalanced datasets

Packt is searching for authors like you

If you're interested in becoming an author for Packt, please visit authors.packtpub.com and apply today. We have worked with thousands of developers and tech professionals, just like you, to help them share their insight with the global tech community. You can make a general application, apply for a specific hot topic that we are recruiting an author for, or submit your own idea.

Share Your Thoughts

Now you've finished *Machine Learning Infrastructure and Best Practices for Software Engineers*, we'd love to hear your thoughts! Scan the QR code below to go straight to the Amazon review page for this book and share your feedback or leave a review on the site that you purchased it from.

https://packt.link/r/1-837-63406-8

Your review is important to us and the tech community and will help us make sure we're delivering excellent quality content.

Download a free PDF copy of this book

Thanks for purchasing this book!

Do you like to read on the go but are unable to carry your print books everywhere?

Is your eBook purchase not compatible with the device of your choice?

Don't worry, now with every Packt book you get a DRM-free PDF version of that book at no cost.

Read anywhere, any place, on any device. Search, copy, and paste code from your favorite technical books directly into your application.

The perks don't stop there, you can get exclusive access to discounts, newsletters, and great free content in your inbox daily

Follow these simple steps to get the benefits:

1. Scan the QR code or visit the link below

https://packt.link/free-ebook/978-1-83763-406-4

2. Submit your proof of purchase
3. That's it! We'll send your free PDF and other benefits to your email directly